A TRAILS BOOKS GUIDE

MINNESOTA UNDERGROUND

& THE BEST OF THE BLACK HILLS

A GUIDE TO MINES, SINKS, CAVES, AND DISAPPEARING STREAMS

DORIS GREEN

-Doris Green

TRAILS BOOKS
Black Earth, Wisconsin

Library of Congress Catalog Card Number: 2002114604
ISBN: 1-931599-24-6

Editor: Stan Stoga
Production: Sarah White
Photos: Doris Green, except where noted
Cover Designer: Kathie Campbell

Printed in the United States of America by McNaughton and Gunn.

08 07 06 05 04 03 6 5 4 3 2 1

TRAILS BOOKS
A Division of Trails Media Group, Inc.
P.O. Box 317 • Black Earth, WI 53515
(800) 236-8088 • e-mail: books@wistrails.com
www.trailsbooks.com

To Michael,

for sharing the journey.

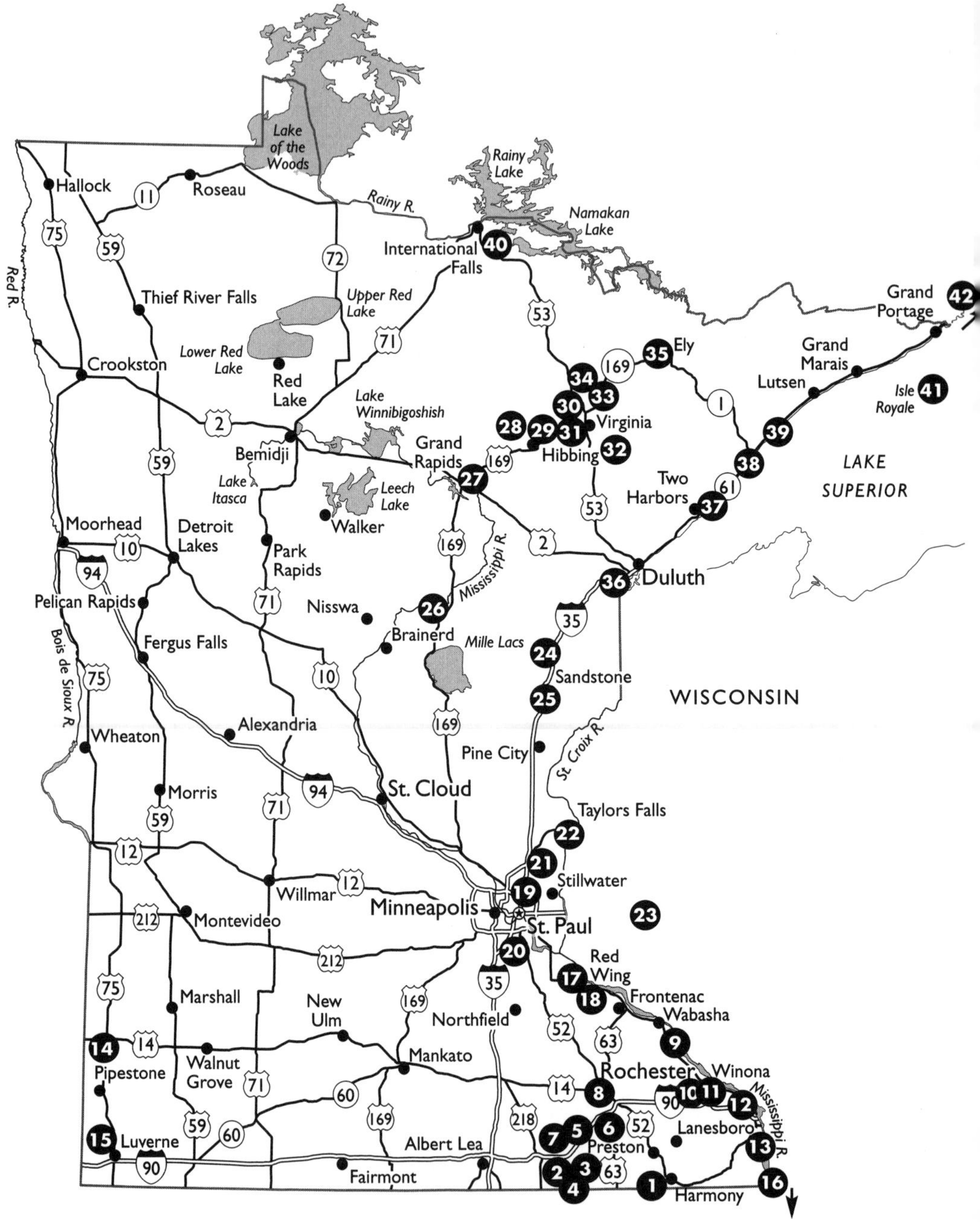
Lake of the Woods
Rainy Lake
Rainy R.
Namakan Lake
Hallock
Roseau
International Falls
Red R.
Thief River Falls
Upper Red Lake
Lower Red Lake
Red Lake
Grand Portage
Grand Marais
Lutsen
Ely
Crookston
Lake Winnibigoshish
Virginia
Hibbing
Isle Royale
Bemidji
Grand Rapids
LAKE SUPERIOR
Lake Itasca
Leech Lake
Two Harbors
Walker
Moorhead
Detroit Lakes
Park Rapids
Duluth
Mississippi R.
Pelican Rapids
Nisswa
Brainerd
Mille Lacs
Fergus Falls
Bois de Sioux R.
Sandstone
WISCONSIN
Alexandria
Wheaton
Pine City
St. Croix R.
Morris
St. Cloud
Taylors Falls
Stillwater
Willmar
Minneapolis
St. Paul
Montevideo
Red Wing
Frontenac
Wabasha
Marshall
New Ulm
Northfield
Rochester
Winona
Pipestone
Walnut Grove
Mankato
Lanesboro
Preston
Luverne
Albert Lea
Fairmont
Harmony
Mississippi R.

Contents

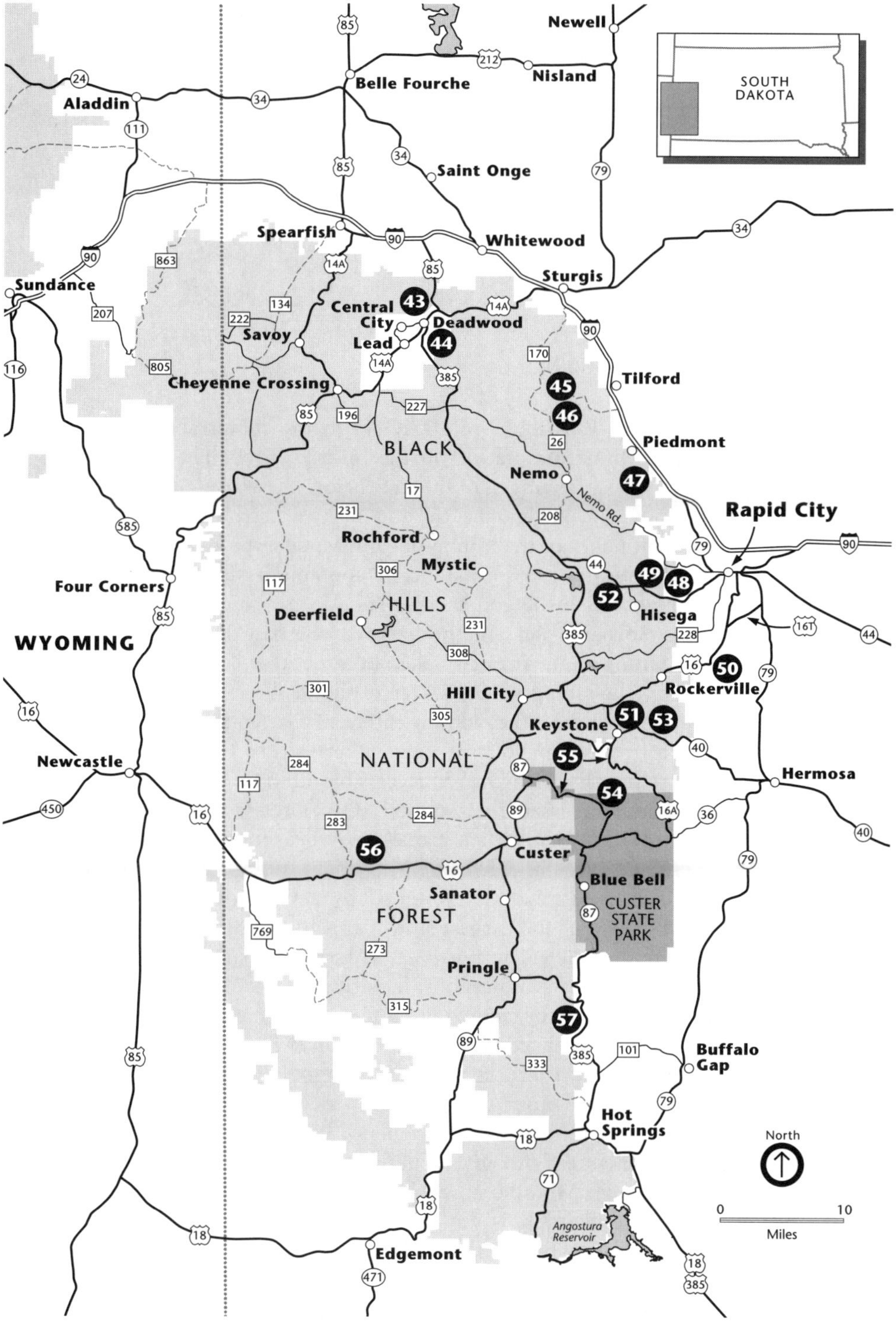

Newell
Belle Fourche
Nisland
Aladdin
SOUTH DAKOTA
Saint Onge
Spearfish
Whitewood
Sundance
Sturgis
Central City
Deadwood
Lead
Savoy
Cheyenne Crossing
Tilford
Piedmont
BLACK
Nemo
Nemo Rd.
Rapid City
Rochford
Mystic
Four Corners
HILLS
Deerfield
Hisega
WYOMING
Rockerville
Hill City
Keystone
Newcastle
NATIONAL
Hermosa
Custer
Blue Bell
Sanator
CUSTER STATE PARK
FOREST
Pringle
Buffalo Gap
Hot Springs
North
0
10
Miles
Angostura Reservoir
Edgemont

Foreword

by Greg Brick, M.S.

THE EARLIEST ALLUSIONS to the subterranean aspects of Minnesota are lost in the mists of time. The Dakotas, who inhabited much of the state before the arrival of European settlers, reverenced Unktahe, the god of the waters. Morgan's Bluff, near the confluence of the Minnesota and Mississippi Rivers, is still said to be the residence of Unktahe. It contains a subterranean passage through which the god travels. The idea of going underground for mere mortals began with the explorer Jonathan Carver. Carver's Cave, containing a spring-fed lake, became the "baptismal font" of Minnesota caving when Carver explored it in 1766 and 1767. His written account is the oldest mention of a cave in the Upper Midwest north of Cave-In-Rock, at the southern tip of Illinois. Carver's discovery now lies buried and inaccessible amid St. Paul's downtown.

Historians assert that Fountain Cave's rocky womb gave birth to the city of St. Paul when a one-eyed voyageur opened a grog shop there in 1838—in which case Minnesota could boast the distinction of having a capital city founded by a troglodyte. Fountain Cave became one of the earliest show caves in the Midwest, giving tours back in the 1850s. The guides carried birch-bark torches for illumination. In nearby Minneapolis, historic Chute's Cave, then known as Nesmith's Cave, began giving subterranean boat tours a decade later. In the twentieth century, with the advent of the automobile, the rugged southeastern corner of Minnesota, dubbed "Little Switzerland," became more accessible, and show caves sprang up. Niagara Cave, named for its waterfall, opened its doors in 1934, and Mystery Cave, the largest known cave in the state, in the 1940s.

Although Minnesota, with its limited area of comparatively thin carbonate bedrock, is not considered a major cave region, there is good reason to believe that its largest cave hasn't yet been discovered. Lengthy groundwater dye-traces from York Blind Valley to Odessa Spring—the largest spring in Minnesota—lead geologists to infer the existence of an enormous cave there. The words of Luther Chamberlin, nom de plume of a Minnesota settler, written in 1867, remain true today: "Minnesota is yet in its infancy, and there is no telling what may yet be found underneath her surface."

This book is the first to deal with Minnesota's underground attractions as a whole (pun intended). As Doris Green points out, the Gopher State has two distinct subterranean regions: the northern half, with its mines, and the southern half, with its caves. Guidebooks to Minnesota caves that have appeared in the past exclude mines from consideration, and vice versa. One of the reasons this book is so useful is that it covers both types of underground sites.

In South Dakota, the story of the spectacular Black Hills caves is much different. The state boasts of some of the longer and more magnificent caves in the world. Native Americans, who regarded the breathing holes and healing waters of this great hunting ground as sacred, were able to exclude white settlers from the area until the gold rush precipitated by General George Custer's 1874 expedition. The town of Hot Springs emerged as a health resort, and nearby Wind Cave began giving tours in 1890. Wind Cave, together with Crystal (now Bethlehem) and Jewel caves, which were commercialized soon after, became known as "the Three Great Caves of the Hills." A half-dozen additional show caves sprang up during the 1920s and 1930s.

Given South Dakota's spectacular caves and historic mines, it's surprising that there is no guidebook to them in print. Travel guides for the state provide scant information on publicly accessible caves, mines, and tunnels. This book will fill that gap nicely.

I hope it will also spark a deeper interest in caving among its readers. If so they should consider joining the National Speleological Society (NSS) or one of its local chapters, called "grottoes." Membership is open to all, including the proverbial armchair caver. I am a geologist by profession and have been a member of the NSS for years.

Why not join the fun!

Greg Brick has a Master's Degree in Geology from the University of Connecticut and is a Professional Associate at an environmental consulting firm in St. Paul, Minnesota. He is a long-time member of the National Speleological Society, Minnesota Speleological Survey, and American Spelean History Association. The author of numerous technical publications about caves, he has traveled nearly 10,000 miles in the preparation of Iowa Underground, *to be published by Trails Books in 2004.*

Acknowledgments

A BOOK IS A team project. It depends on the expertise and generosity of many individuals who aim to set a straight record—in this case about the many underground mazes that I write about here.

This guide would be less true, less accurate, and less complete without the valuable contributions of Greg Brick, the geologist who penned the book's foreword and provided information on many of the sites described.

Other contributors provided site background, shared caving and mining knowledge, and offered encouragement:

- John Ackerman shared the story of his development of the Cave Farm Karst Preserve.
- E. Calvin Alexander, Jr., Morris-Alumni Professor, Department of Geology and Geophysics, University of Minnesota—Twin Cities, provided many helpful comments on the manuscript.
- Donna Bremer, co-owner of Wabasha Street Caves and the Joseph Wolf Caves, offered recent historical research on the Stillwater caves.
- Blaze Cunningham, co-owner of Crystal Cave, provided updated information on the cave's development and preservation.
- Dave Gerboth offered useful information on Odessa Spring.
- Randy D. Gordon, park manager at Banning State Park, provided a photo and background on Robinson's Ice Cave.
- Mike Jorgenson discussed the creation of Stone Creek Park.
- Steve Lukinuk, co-owner, Amethyst Mine Panorama, provided a lesson in amethyst mining and processing.
- Bill Majewski, a Duluth city planner, and Nancy Nelson explained the history of Ely's Tunnel.
- Grant W. Nelson, Sr. shared his boyhood explorations of the caves in Winona's bluffs.

- Warren Netherton, Department of Natural Resources cave specialist at Mystery Cave, pointed out features and resources I might have otherwise missed.
- Mary Lou Santovec gamely climbed rocks, scaled bluffs, and slid down ravines with me in search of caves in southeastern Minnesota.
- Gary Soule, Wisconsin cave historian, generously shared his expertise and library.
- Joe Terwilliger welcomed me to a Minnesota Speleological Survey invitational weekend.

Many other individuals directed me to underground sites, answered my questions, and shared their stories of Minnesota's underground.

Stan Stoga, acquisitions editor at Trails Books, provided expert editing and shepherded the final manuscript into a finished book, ready for your bookshelf, reading table, nightstand, or caving weekend. ■

Introduction

CATS CAN SENSE when an earthquake is imminent long before humans. In fact, many of us fail to realize the hidden wonders, natural and even man-made, that lie beneath our feet. We walk down a city sidewalk and see cement—not giving a thought to the water and sewer tunnels, power and communications cables, old building foundations, even caves that may lie below. We drive through rural areas, flat fields as well as rolling hills, without contemplating the undiscovered natural treasures that may lie just below the surface. We drop a line in a lake, not realizing that it is a reclaimed pit mine.

Such is the case with Minnesota. Some of the state's most intriguing attractions lie just beneath its surface, in the ore outcrop of a pit mine, the cool shade of a rock-cut tunnel, or the shelter of a sandstone ledge. With more than a dozen miles of passages, the Mystery Cave system is one of the largest in the Upper Midwest. Northern Minnesota's iron mines annually produce enough taconite pellets to fill half-a-million railroad cars, and the area's prodigious iron output was a factor in the Allied victory in World War II. Taconite mining continues to play a significant role in the state's economy. The Cold Spring Granite Quarry in Rockville has produced countless tons of high-quality pink and gray stone since production began in 1889. And the underground storage cave at Rochester's Quarry Hill Nature Center surely is one of the largest in the Upper Mississippi Valley.

Neighboring South Dakota is a different story entirely. For many years, the Black Hills have attracted tourists to its rich array of underground wonders. Its marvelous caves contain a multitude of speleothems, including some crystal formations seldom found in Minnesota. Jewel Cave and Wind Cave are among the longest in the world, attracting tens of thousands of visitors annually. Seven other commercial caves offer a variety of tours—from easy walks to rugged climbs. South Dakota also has a few accessible gold mines that provided the impetus for a brief gold rush in the nineteenth century. The Black Hills also offers several man-made tunnels that provide an insider's view of the ancient rocks forming the base of this range.

This guidebook features 57 publicly accessible underground and in-ground attractions, as well as replicas that give the visitor the feeling of being beneath the surface. First and foremost, the book takes readers on a tour of Minnesota's rich variety of caves, mines, and tunnels, many of which are relatively unknown and underappreciated. The book also provides profiles of 15 South Dakota underground attractions, a couple of them justifiably world famous, others less well known but still worthy of exploration by adventuresome travelers. My hope is that the book will pique your interest enough to enjoy an extraordinary underground adventure in both states.

A Long-Range View

Imagine planet Earth in its infancy, a cooling mass of molten rock, spinning in orbit around the sun. During the **Precambrian** eon—more than 570 million years ago—primitive one-celled and multicelled organisms first appeared. Abundant sea life and the appearance of the first vertebrates and land animals marked the early stages of our own eon, the **Phanerozoic**—when life became widespread. This eon is divided into three eras:

- The **Paleozoic**—which featured a variety of early life forms and ended about 230 million years ago. Shallow seas covered much of the state during this era.
- The **Mesozoic**—which included the era when dinosaurs dominated the planet and ended about 65 million years ago. In Minnesota, lands uplifted, and wind and water eroded the exposed rock.
- The **Cenozoic**—the current era in which mammals have dominated the earth.

In turn, geologists divide each of these eras into distinct periods. For example, many Minnesota caves are located in rock layers formed during the **Ordovician** period of the Paleozoic era.

These periods are further subdivided into epochs, according to what significant events occurred. For instance, in the **Pleistocene** epoch of the most recent period of the Cenozoic era (the **Quaternary**), a series of glaciers up to a mile thick invaded most of Minnesota. This Ice Age began about two million years ago and lasted until approximately 10,000 years ago. As the glaciers repeatedly pushed forward, then melted and retreated, they left depressions, valleys carved by rushing meltwater, and mounds and ridges of boulders once caught up in the ice.

Except in southeast Minnesota. This area, along with southwest Wisconsin and northeast Iowa, was largely untouched by glaciers. Free of glacial debris and lakes, the area was primarily shaped by rivers that once drained melting glacial waters. **Limestone**—composed chiefly of calcite—and **dolostone**—composed chiefly of dolomite—dating from the Paleozoic era lie near the surface. These layers are very soluble, and over thousands of years weak carbonic acid, created from rain mixing with carbon dioxide, have carved multiple channels and passages below ground. Some are explored in more detail in section I.

The caves in the Black Hills, explored in section IV, developed in limestone generally overlaid by a layer of sandstone. Yet granite and other Precambrian rock is also visible here, because the sandstone and limestone layers have eroded and entirely disappeared from the high, central core of the hills. Precambrian layers can be seen in the needles formations in Custer State Park. Similar outcrops, tunnels, and a few former gold mines now are accessible to the public.

How the Book Is Organized

The book is divided in four parts along broad geographical divisions: south, central, and northern Minnesota, and the Black Hills of South Dakota. While each area features certain predominant geologic characteristics, there is great diversity as well.

1. The southern part of Minnesota includes the caves, rivers, and springs of the **driftless**—or unglaciated—area. But it also includes the quarry at Pipestone National Monument in the southwest part of the state.
2. Minnesota's central section features the sandstone caves of St. Paul, the granite of St. Cloud, and the potholes of the St. Croix River Valley at Interstate State Park.
3. The northern region contains the underground and openpit mines of the Mesabi Range, and a few tunnels and several wonderful sea caves along the Lake Superior shore.
4. The section on the Black Hills should convince you that the trip across South Dakota is worth the effort. The area is home to more than a dozen large and publicly accessible caves, mines, and tunnels. There are tunnels constructed on highways and railroad lines, as well as three former gold mines. In addition, nine significant caves in a limestone belt circling the center core of the Hills are open to the public.

The book also contains two sites just over the border from Minnesota in Wisconsin and one on Isle Royale (which is a part of

Michigan but located much closer to Minnesota). They seem almost an extension of Minnesota geology. By contrast, another site in the book, the Amethyst Mine in Thunder Bay, Ontario, seems a world away in terrain, though only a few hours drive from the Minnesota border.

In addition, many of the sidebars describe sites that may not qualify as full-fledged underground attractions but that should fascinate anyone interested in exploring a range of geologic wonders.

Each of the 57 sites in the book includes driving directions and most specify seasons and hours of operation, amenities, and precautions. The sites differ in accessibility. Some can be traversed by wheelchair; others can be reached only via a challenging hike.

Safety, Safety, Safety

All sites listed here are publicly accessible and reasonably safe. Several involve moderately difficult hikes, and a few are very challenging. The best precaution is to know your skills, experience, and limitations—not to mention those of accompanying children and older adults.

This book is designed as an introduction to underground and in-ground attractions in Minnesota, and in South Dakota. Learning about the state's geology and underground history can deepen your understanding and appreciation of its rivers, springs, lakes, and other natural features. The book can point out intriguing aboveground features and direct you to guided underground tours. It can increase your knowledge of the state's down-under world.

But reading this book is not a substitute for hands-on training in cave exploration. If, after exploring some of the sites listed here, you want to go to the next level and explore more challenging caves, it is time to connect with a local speleological group that can guide you safely into the underground world. The Minnesota Speleological Survey (see page 166) is one such organization. It and groups like it know where to go safely, how to prevent mishaps, and what to do in the event a problem arises.

When it comes to caving, accidents can happen quickly, and they can be deadly. Ask the family of the college student who didn't make it out of a Minnesota cave when it suddenly flooded. More recently, a father, his teenage son, and a friend crossed the border into Iowa and entered a cave clearly visible from the highway in a road-cut. Two hundred yards from the entrance, one of the teens tried to push through a narrow passage and got stuck. He was wedged there for 12 hours before experienced cave rescuers managed to free him. Twelve hours immobilized by rock in an environment where the average temperature is 47 degrees Fahrenheit presents a threat of hypothermia, as well as other risks.

Never enter an abandoned mine. Cave-ins are a real threat. Loose boards with nails, rocks, and rattlesnakes are among the other hazards of any mine. When touring a commercial mine, you may want to bring your own hardhat (most but not all commercial mine operators supply hardhats to visitors) and a flashlight, as passages are usually dimly lit. Wear lace-up shoes with patterned rubber soles. Wear old clothing, since the heavily mineralized mud in many mines can permanently stain leather and fabric. While commercial mine operators often permit young children to tour a mine with adults, mine experts sometimes caution against taking children under the age of 12 into any mine. Certainly, you should keep a careful eye on children and prepare them in advance for the experience, impressing upon them the importance of staying close and not picking up or touching boards, equipment, or other items.

The bottom line: don't explore an unknown underground passage without an experienced guide. The commercial tours listed in this book showcase many breathtaking underground features, and other sites offer grand views, solid information, and historical lore. But when you are ready for more, join a group.

Interacting with the Underground Environment

When you visit the underground and in-ground attractions listed in this guidebook, you'll see many signs of human impact, positive and negative. Mining reclamation efforts have converted former pit mines into beautiful lakes used for scuba diving, fishing, swimming, and other water sports. The mines themselves have spurred local economic development and provided jobs for thousands of workers over the years. At the same time, careless mine owners have made mistakes that cost lives. Wetlands important to wildlife have disappeared under urban development. And, pollutants have leached into the passages and cavities that underlie the southeastern corner of the state.

Your choices and actions make a difference in the state's ecology. Do you toss a cigarette butt onto a cave floor, or do you pick up a soft-drink can along a hiking trail? Do you pocket a prize fossil—the remains of an ancient leaf, snail, or other organism preserved in rock—that you picked up from an area where fossil hunting is prohibited? Or do you point it out to your companion, photograph it, and leave it where you found it? Especially if you travel under the watchful eye of children, your actions may have a long-term impact on the environment.

Many cavers follow this credo, promoted by the National Speleological Society: "Take nothing but pictures. Leave nothing but footprints. Kill nothing but time." It's good advice for the attractions in this book. ■

Section I
Southern Minnesota

DRIVE FROM WEST to east almost anywhere south of the Minnesota River and the landscape changes dramatically—from generally level plains to rough countryside marked by steep stream valleys, rugged bluffs, sudden depressions, and cone-shaped pits. The last glaciers of the most recent Ice Age advanced across the southwestern part of the state while leaving the Southeast untouched. More than 10,000 years later, this demarcation shows above and beneath the surface of southern Minnesota.

When the last glacier retreated from southwest Minnesota, it left a thick layer of soil, stones, rock, and other debris, called **till**. Eventually the glacial till supported fir and spruce forests. Over thousands of years these forests gave way to the vast prairies that European explorers found here. In the nineteenth century, pioneer farmers cultivated this region, which today accounts for the major part of the state's grain production.

Depending on your route, you may encounter a highland in Minnesota's southwest corner, edged with a cliff—called an **escarpment** or **scarp**—visible in sections of the east-facing side. This is Coteau des Prairies. One meaning of the French word coteau is "knife." This knife-edged upland separates the James River and Minnesota River basins and is the Mississippi River–Missouri River watershed.

These high prairies contain outcrops of Precambrian Sioux quartzite, visible at Pipestone National Monument and Blue Mounds State Park. The 1.5-billion-year-old quartzite far predates the glaciers. Usually pink or red in color, this quartzite contains thin layers of pipestone, quarried by Native Americans for use in pipes and other objects. (See page 43.)

After you leave the Coteau des Prairies behind and cross miles of broad, monotonous fields, you reach the Owatonna **Moraine**—a hilly area of glacial debris, running north and south through Albert Lea. This moraine marks the edge of the last glacier advance, before it began melting and retreating to the west and north.

East of the moraine, the Precambrian layers are deeply buried beneath Paleozoic rock. Glaciers marked the landscape more by their retreat than by their advance: rushing glacial meltwater carved deep river gorges and underground stream channels. Layers of limestone and dolostone continue to be undercut today by eroding waterways and groundwater seepage.

Throughout southeast Minnesota, rainwater seeps through cracks and fissures in the limestone and dolostone. The rainwater combines with carbon dioxide gas in the soil, creating a weak carbonic acid, which dissolves the stone. Over thousands of years, this acid creates and enlarges cavities in the limestone and dolostone layers, often along joints in the rock. This is one way in which caves are formed.

Often limestone and dolostone **solution caves** contain fossils and formations created when the dripping acid leaves behind calcite and other mineral deposits in the shape of **stalactites,** formations similar to icicles growing down from the ceiling; **stalagmites,** cones growing up from the floor; **columns,** where stalactites and stalagmites meet; and other **speleothems** of various shapes and sizes. Most of the caves in southeastern Minnesota are in limestone and dolostone created 400 million years ago during the Ordovician period.

The points where groundwater enters fissures and cracks sometimes erode and widen into **sinkholes** or **sinks**—funnel-shaped openings in the rock layers. Sinkholes may also develop rather suddenly, for example, in a collapse during road or building construction or near man-made ponds or sewage lagoons. Sinkholes vary in size, from a few feet to as large as 100 feet in diameter and up to 30 feet deep.

Driving through southeast Minnesota, you can spot sinks in fields. They typically appear as depressions in the earth, sprouting trees and brush in contrast to the surrounding agricultural crops or pasture. You also can see a sinkhole near the welcome sign to the town of Fountain on U.S. Highway 52. Fountain, in fact, bills itself as the sinkhole capital of the United States.

Sinkholes sometimes form when part of a cave system beneath them collapses. This can occur when the groundwater table in the area is lowered and water drains from a previously water-filled cavern, leading to the collapse of the cavern roof. Sinks also form at points where joint planes in underground rock layers intersect, providing vertical drainage passages for rainfall and other surface water.

While sinkholes can extend as deep as five hundred feet, Minnesota sinks are usually less than fifty feet in depth. Many lead to underground cave systems or streams. Minnesota cavers have sometimes found new caves or extensions of known cave systems by digging into the bottom of sinks.

Until the latter half of the twentieth century, farmers and other landowners often viewed sinks as a nuisance and sometimes filled them with trash, rocks, or other debris. People were sometimes careless about allowing agricultural runoff to enter sinks, polluting the underground water supply. In more recent years, public awareness about the susceptibility of the water supply to pollution has grown and some sinks have been cleaned out.

The landscape of southeast Minnesota is ever changing: new sinks open up. Existing sinks may slowly be filled in, as runoff carries sediments that build up in the bottom of the sinks. Over decades, sinks may become recognizable only as clumps of trees growing in a depression in a field. Southeastern Minnesota sinks are numerous. Fillmore County alone may contain as many as 10,000.

This Swiss-cheese landscape, filled with sinks, fissures, underground stream channels, and caves, is called **karst** or **karst topography**. Named for a limestone plateau near Trieste, Italy, karst describes 20 percent of Minnesota's landforms, as well as 20 percent of U.S. topography. Major karst areas in the United States include Tongass National Forest in Alaska, central Kentucky, southeastern New Mexico, the Missouri Ozarks, and western Texas.

While most of southeastern Minnesota exhibits karst, it is most developed in Fillmore County, which contains an abundance of sinks, caves, and streams. The 13-mile-long Mystery Cave system is in Fillmore County, as is the 5-mile Spring Valley Caverns. Most caves are, of course, much smaller, containing only a few passages or rooms. Generally, a cave is defined as any underground cavity large enough for an adult to enter.

Southeastern Minnesota streams and rivers lie above ground, underground, or frequently both. A stream might disappear below the earth at the end of a **blind valley**—a valley with a dead end, similar to a box canyon—only to reappear as a spring at the base of a river bluff. Occasionally the spring opening has eroded into a cave that may extend into the bluff for some distance.

The underground labyrinth is a complex system of channels, passages, caves, and fissures. Scientists have conducted dye studies to determine how quickly pollution spilled into a stream might travel underground and where it might resurface. Such studies have documented complex paths, with dye from a single spill tracked at the outflow of several springs and at greatly varying rates.

Karst topography is particularly sensitive to pollution. Many other aquifers are composed of sand, usually containing some clay. Water flows much more slowly through sand than it does through limestone, giving time for pollutants to be filtered out and broken down—either by bacteria or through simple breakdown into nontoxic components. In addition, the clay minerals in most sands absorb heavy minerals and some other pollutants. Understanding and exploring Minnesota's karst can enable us to better protect this fragile resource. ■

1

Niagara Cave
Fillmore County

APPROACHING THE entrance building from the parking lot at Niagara Cave, the first thing you notice is a slanting trough of running water. Simulated "Sandy Creek" is a sluice where children and adults can realize their dreams of "panning" (screening) for gemstones. You can purchase a bag of mining "rough," place it in a screen-bottomed wooden box, dip it in the sluice, and watch your gems appear as the soil washes away.

The real gem, however, is beneath your feet. Niagara Cave plunges underground along a series of fractures in a cross-section pattern, with "section" lines at approximate 45-degree angles. The tour follows a zigzag path through these eroded fractures. In some passages, the ceiling soars far above your head; in other places, you walk over a grate and can see the cave floor scores of feet below. As the tour snakes downward through narrow gorges and canyons—some no more than 4 feet wide—you may feel as if you are walking through the eye of a very large needle.

Descending, you soon hear a muffled thunder from somewhere ahead. Your guide stops to describe the 60-foot waterfall ahead, since it would be useless for him to try to describe the feature when you finally stand looking over the precipice in the midst of spray and a crescendo of cascading water. Sixty feet below, the water churns in chaos; 70 feet above, the domed ceiling shelters the echoing tumult.

The waterfall and a wishing well are down the opposite arms of a short side path. In addition to the standard cave wishing well (this one is water-filled pothole scoured out by a whirlpool), Niagara provides another make-a-wish opportunity: the legend of Kissing Rock (a limestone shelf that fell from the ceiling) is that if you kiss someone on this table, the person you kissed will be yours forever.

As Niagara Falls is a traditional honeymoon destination, so romance abounds at namesake Niagara Cave. More than 300 couples have been married in the Crystal Wedding Chapel, which features a white lectern, benches, an archway, and natural decorations. Soda straw stalactites hang from the ceiling in the rear of the chapel; "stained glass" windows, ribbon stalactites (cave bacon), and flowstone decorate the walls. A bit farther along the tour is another romantic favorite: the Bridal Veil is a 75-foot cascade of ribbons and flowstone.

This stream in Niagara Cave cascades from a 60-foot waterfall into a 130-foot dome.

Every visitor has at least one favorite formation. Other candidates:

- The Wedding Cake has three layers and flowstone frosting.
- A skeletal hand and arm—complete with radius and ulna—protrude from high on the wall near the Kissing Rock.
- The Stalactite Room features numerous huge limestone icicles, thousands of years old, including sharks' teeth stalactites, which drop down along the wall.
- The Liberty Bell—complete with crack—was formed by an ancient whirlpool.
- Paul Bunyan's huge bed was eroded by rushing water.
- The Elephant's Head is composed of dried-up flowstone and is no longer growing.
- The Battleship is an aptly named lump of limestone.

In many places throughout the cave, you can find various fossils. Several examples of sunflower coral, or *Receptaculites*, exist high on the cave walls. This small organism once lived in the ancient seas that covered southeastern Minnesota. A cross-section has the appearance of a flower. A cephalopod visible near eye level also catches attention.

Here and there, lumps of chert stick out of the wall, where other rock has eroded around the hard substance. Niagara Cave plunges

Odessa Spring

Southeastern Minnesota's largest spring empties into the Upper Iowa River, flowing from a cave in a Minnesota bluff east of Granger in Fillmore County near the Iowa state line. Odessa Cave is on private land but is visible from a canoe in the river. The opening is large and the cave extends roughly 50 feet into the limestone bluff.

On its journey toward the river, the stream runs underground for more than ten miles, descending through several rock layers. It sinks underground at the end of York Blind Valley after traveling aboveground across surface glacial deposits. When the stream reaches the dead end of the blind valley, it sinks into the karst system of Fillmore County—and then pops up again at Odessa Spring.

Another large spring is at the Department of Natural Resource's Lanesboro Fish Hatchery. It is the resurgence of Duschee Creek.

Springs in the karst of southeastern Minnesota often begin as aboveground streams, vulnerable to various types of pollutants. Underground streams in this region may also become polluted from materials entering them through sinkholes. The term "pure spring water" makes great advertising copy, but in karst country it's usually a misnomer. When you encounter a clear spring in southeastern Minnesota, don't drink the water.

through three distinct limestone layers, with the Dubuque Formation lying nearest to the surface. Underneath it lie the Stewartville and the Prosser Members of the Galena Formation. All three layers were formed when seas covered this area 400 million years ago.

At the deepest point in the tour, you are more than 200 feet beneath the surface. Many more rooms are accessible only with scuba gear. The main branch of the river that runs through the cave comes to the surface three miles to the south at Hawkeye Spring in Iowa.

Current owner of Niagara Cave Mark Bishop continues to clear the lower passages. People have been digging out and exploring the cave since it was discovered in 1924 by two farm boys. They followed three lost pigs into a hole in the earth. The boys heard the pigs squealing and followed them to a ledge 75 feet below the surface. After the boys used a rope to rescue the pigs, they returned to explore the cave. More than three-quarters of a century later, adventurers continue to visit Niagara's treasures.

One feature modern-day visitors don't see is a bat of any type. Because the entrance building covers the only opening to the cave, bats are unable to gain admittance to Niagara.

Directions: From Rochester, follow U.S. Highway 52 to the town of Harmony. In town, take State Highway 139 south and drive 2.5 miles to County Highway 30. Turn right and proceed 2.5 miles to the cave. Niagara Cave is less than an hour drive from Rochester and from exits off I-90 in Austin and Winona.

Season/hours: Open weekends in April, 10 a.m. to 4:30 p.m.; May through September, 9:30 a.m. to 5:30 p.m.; weekends in October, 10 a.m. to 4:30 p.m.; open in the off season by appointment. Fee.

Length: Tour lasts about one hour and covers approximately three-quarters of a mile.

Precautions: Five hundred steps make this tour a challenge for some individuals. Walking shoes and a light jacket are recommended; temperature is 48 degrees year-round.

Amenities: Picnic tables, playground equipment, restrooms, and gift shop.

Contact: Niagara Cave, P.O. Box 444, Harmony MN 55939. Phone: (800) 837-6606. E-mail: niagara@polar istel.net. Web site: www.niagaracave.com.

2
Mystery Cave I
Fillmore County

ENCOMPASSING ABOUT 13 miles of known passages, the Mystery Cave complex is the largest in the Upper Mississippi Valley. The complex lies within a satellite section of the Forestville/Mystery State Park and features a maze of linear passages in three predominantly limestone layers (Dubuque Formation, the Stewartville Member of the Galena Formation, and the Maquoketa Formation).

Millions of years ago, the solution cave system began forming along joints in the layers as the weak carbonic acid of rainwater seeped downward, dissolving the limestone. In some places today you can see fossils of brachiopods, cephalopods, and other early sea creatures embedded in the ancient limestone. Thousands of years ago, stalactites, stalagmites, and other multihued formations began developing in the underground cavities. Stalactites and stalagmites are formed when calcium carbonate precipitates from a solution that has become supersaturated as the result of evaporation, an increase in temperature, or a change in the water chemistry.

Mystery's caverns offer adventure for children and adults of all ages. A one-hour commercial tour of Mystery I is wheelchair accessible, and a two-hour tour of Mystery II (see page 18) provides an opportunity to see a less developed section of the system. Experienced cavers and researchers can receive permission to explore, survey, and study wilder sections of the cave network.

The Mystery Cave complex was not discovered all at once. After World War I, Mystery Cave Park attracted visitors to Old Mystery Cave and the wooded plain along the Root River. By the mid-1930s visitors could pay a fee to tour the 650-foot main passage, but a devastating flood put an end to the commercial venture in 1942. Today Old Mystery Cave, one of several parts of the Mystery Cave complex, lies on private land outside the state park and is inaccessible to casual visitors.

Joseph Petty discovered Mystery I (then called Mystery Cave) in 1937, after observing melting snow and steam near crevices in the rock surrounding the entrance to an underground stream. He explored the cave with property owner Walter Blakeslee, who in 1938 began offering a one-hour tour similar to the tour offered today.

The Mystery Cave system contains many formations, 13 miles of passages on two levels, and several pools.

Mystery I was closed during World War II. After the war, new owner Clarence Prohaska improved the cave and opened it for tours in 1948.

You enter Mystery I through a large steel door anchored in the side of the bluff above the South Branch of the Root River. Tiny windows in the door permit entry and egress to the four species of cave bats that hibernate in the cave. The entrance is in the Dubuque Formation, featuring olive-gray limestone with interbedded shale. The northeast trending passage is an easy walk: it's paved, well-lit, 10 to 20 feet wide and up to 30 feet high.

The Mystery I tour offers the following memorable features:

- The Cathedral Room features a chapel with a 30-foot-high vaulted ceiling reaching up to the Maquoketa Formation, another distinct limestone layer.
- The Stalactite Room offers a variety of other slowly growing formations, in addition to the rock icicles hanging from the ceiling.
- Turquoise Lake is a clear pool ten feet deep with a flowstone bottom, stalactites, and cave popcorn concretions along one side.
- The Bomb Shelter is the final room of this tour.

In addition to several natural and artificial pools, the Mystery Cave system contains several underground streams, including a section of the South Branch of the Root River that cuts across a looping meander in the river. Cooled by its sojourn underground, the river downstream of the cave supports a healthy trout population. There is good fishing in Forestville State Park.

Directions: From Rochester, follow U.S. Highway 63 to State Highway 16. Turn left on 16 and drive west through Spring Valley. Turn right on County Highway 5 and drive 4 miles to County Highway 118. Follow 118 and watch for signs to the cave. Mystery Cave is less than an hour drive from Rochester and from exits off I-90 in Austin and Winona.

Season/hours: Open weekends from mid-April to Memorial Day weekend and after Labor Day through October; open daily from Memorial Day weekend through Labor Day weekend. Fee.

Length: The regular tour lasts about one hour.

Precautions: The regular tour is wheelchair accessible and features paved passages. A light jacket is recommended; temperature is 48 degrees year-round.

Amenities: Forestville/Mystery Cave State Park offers picnic tables, playground equipment, restrooms, trout streams, and recreational trails. Also in the park, Historic Forestville is a restored 1800s village operated by the Minnesota Historical Society. If you visit in May, watch for the carpet of Virginia bluebells along the drive to the park office.

Contact: Forestville/Mystery Cave State Park, Route 2, Box 128, Preston, MN 55965. Mystery Cave phone: (507) 937-3251. Web site: http://www.dnr.state.mn.us/parks/forestville_mystery_cave/index.html.

3 Mystery Cave II
Fillmore County

THE TOUR OF Mystery Cave II descends through layers of limestone and shale into dolostone, which is rougher in appearance here, thus offering visitors a graphic lesson on the difference between the two substances. The shapes of the walking passages are also different in the two layers. Passages within the dolostone tend to have rounded or even pointed ceilings along joint planes, while passages in the limestone layers tend to be as flat as a mine tunnel. Often, a limestone layer has fallen from the ceiling, where a softer layer of shale intersperses two limestone layers, creating the flat appearance.

A short walk from the entrance leads to a large, straight passage, 5th Avenue. To the left (east), it leads to the Garden of the Gods, featuring numerous stalactites and other formations on ceiling, walls, and floor. To the right are rooms, side passages, and formations.

One of Mystery Cave II's unmistakable formations are these raft cones protruding from Blue Lake. Courtesy of Minnesota Department of Natural Resources.

Blind Valley Scientific and Natural Area

A few short miles south of Mystery Cave, you can walk aboveground over another cave system. Donated to the state by the Thomas Kapper family, Blind Valley Scientific and Natural Area preserves a 40-acre parcel of karst and offers trails, campgrounds, and picnic areas.

The Department of Natural Resources has gated the entrances to several caves, including Goliath Cave and Woodchuck Cave, for safety. (Some passages are largely water filled, and much of this cave system is prone to flooding during thunderstorms.) But you can still walk aboveground and get a close-up look at sinkholes and a sinking stream at the end of a blind valley.

To reach Blind Valley from Rochester, follow U.S. Highway 63 south to State Highway 16. Turn left on 16 and drive east through Spring Valley to County Highway 5. Turn right on 5 and go its junction with County Highway 14 (1 mile south of the Mystery Cave turnoff on 5). Proceed 1.5 miles south on 5 through Cherry Grove. Turn left (east) on County Highway 20 and drive 1.25 miles. The Scientific and Natural Area is on your left. Turn left and drive less than a half-mile to the parking lot on your left.

Other highlights of the Mystery II tour include:

- The Carrot Sticks are twin stalactites, colored by hydrous iron oxide incorporated into the calcium carbonate layers of the formation.
- The Hills of Rome feature beds of stone rising through the Stewartville and Dubuque layers.
- Blue Lake, the largest pool in Mystery, has a water level that fluctuates during the year, drying up each winter. The lake contains unusual raft cones and calcite rafts. When the water level lowers, the rafts sink onto the cones and build them up. Some of the cones are seven feet high.
- The Smoking Chamber contains an artificial pool constructed beneath a ceiling drip.
- Several **vugs**—these small cavities in the passage walls are lined with crystals, similar to geodes.
- **Helictites**—hook-, twig-, or bushlike formations created when water seeps slowly out of the side of a stalactite instead of dripping straight off the end. Helictites tend to form in drier caves.

Mystery II (also known as Minnesota Caverns) was discovered in 1958, when Mike McDonald pushed through a thinning tunnel in Mystery I. McDonald later returned with a small group of cavers and explored the long extension of this cave system. Mystery II is still accessible from Mystery I by negotiating a 10-hour walk, crawl, and squeeze—but it's much easier to enter via the man-made entrance used by tour groups.

Since then, additional passages and rooms have been discovered and mapped. In 1967, cavers Doug Nelson and Craig Porte discovered Mystery III; other cavers explored the new section, including Lily Pad Lake, Eureka Avenue, and Dragon's Jaw Lake. Each year, cavers explore, clear, and survey more of the far reaches of the Mystery Cave complex.

Directions: Season/hours, amenities, and contact: the same as those for Mystery I.

Length: The Mystery II Tour takes two hours.

Precautions: Expect to carry your own lantern. The trail is occasionally wet and slippery.

4

Forestville State Park

Fillmore County

THE MAIN SECTION of Forestville State Park, which operates Mystery Caves, also contains a few small caves of its own, though they hardly compare to the grand Mystery system. Most of these are off the recreational trails and difficult to find. Big Spring on Canfield Creek, however, is worth the hour-plus trek to the end of a long spit of land that juts south from the rest of the park.

Water from the spring has hollowed out a small cave in the side of a bluff and in summer and fall spreads out across a pebble "beach" before entering the main current of the creek. Wearing boots, you can stand at the entrance of this cave as the water flows over your feet. Though it may appear to issue from the depths of the earth, this stream actually sank underground less than five miles upstream.

Boots make a hike to the spring easier, since it involves two fords across the creek. Solve this problem by riding horseback to the spring. There is a hitching post nearby.

Watch for wildlife along the wooded trail. On one fall visit, a pileated woodpecker flew overhead and three bald eagles soared above the spring and bluffs. You may also spot turkeys, soaring turkey vultures, and scarlet tanagers.

Deer, beaver, raccoon, mink, red and gray fox, and other animals make their home in Forestville State Park, as do a growing number of coyotes. If you have heard only a few coyotes howling at dusk in other

Big Spring in Big Spring Cave at Forestville State Park flows over a shallow pebble beach into Canfield Creek during summer and fall.

locales, you may want to plan your hike to Big Spring in mid afternoon. Returning at dusk, you may hear scores of coyotes yipping, howling, and crying from one bluff to another, their voices amplified in the resounding valley.

There are a couple of other springs in the vicinity. Rainy Day Spring is in Forestville State Park, though there isn't a trail to it. Moth Spring, outside the park at the head of Forestville Creek, is accessed through the private Maple Springs Campground, where you can pay to park your vehicle while you hike. The drive to the campground is near the entrance of the state park.

Directions: From Rochester, follow U.S. Highway 63 to State Highway 16. Turn left on 16 and drive west through Spring Valley. Turn right on County Highway 5 and drive 4 miles to County 118. Follow 118 about 2 miles to the park. Forestville State Park is less than an hour drive from Rochester and from exits off I-90 in Austin and Winona.

Season/hours: Open year-round, however, some facilities (shower building, campgrounds with electricity, campgrounds for horses) and trails close in the winter. Fee.

Length: The walk from the anglers' parking lot to Big Spring is 3.5 miles.

Precautions: Big Spring is not visible during the springtime and other periods of high water. Since the spring is on public hunting land, wear blaze orange during hunting season. The trail to the spring includes two fords over Canfield Creek—easy on horseback, somewhat challenging on foot. Waterproof footgear is recommended.

Amenities: Forestville/Mystery Cave State Park offers picnic tables, playground equipment, restrooms, trout streams, and recreational trails. Also in the park, Historic Forestville is a restored 1800s village operated by the Minnesota Historical Society.

Contact: Forestville/Mystery Cave State Park, Route 2, Box 128, Preston, MN 55965. Mystery Cave phone: (507) 937-3251. Web site: http://www.dnr.state.mn.us/parks/forestville_mystery_cave/index.html.

Eagle Bluff Environmental Learning Center

People of all ages come to Eagle Bluff Environmental Learning Center, southeast of Rochester, for educational programs, meetings, retreats, camps, and even weddings. Students can learn outdoor skills, along with information about wildlife, forests, and geology—particularly the karst of southeast Minnesota, with the aid of a karst exhibit built by the Science Museum of Minnesota. Though designed as a traveling exhibit, the display's permanent home is at Eagle Bluff.

Eagle Bluff also has a tunnel, one of several in the state constructed for hydropower projects. This one on the North Branch of the Root River was completed in 1915. The river looped around a hill, descending 20 feet in the 3.5-mile hairpin bend. Brightsdale Tunnel closed this loop, and the water that cascaded through it turned the turbines in the Root River Power and Light Company powerhouse. This plant provided electricity to several hundred customers until 1928, when it could no longer keep up with the increasing demand. You can still see the remains of the powerhouse at Eagle Bluff. The Brightsdale Tunnel is closed to humans and preserved as a sanctuary for hibernating bats.

To reach the center from Rochester, follow U.S. Highway 52 to Fountain. Turn left on State Highway 8 and drive 7 miles to County Highway 21; turn left and watch for signs.

Contact the Eagle Bluff Environmental Learning Center, 1991 Brightsdale Road, Route 2, Box 156A, Lanesboro, MN 55949. Phone: (888) 800-9558. Web site: www.eagle-bluff.org.

5

Cave Farm Karst Preserve
Fillmore County

WHERE MOST PEOPLE fear to tread, John Ackerman has eagerly ventured—and Minnesota cavers are glad of it. Since 1990, Ackerman has purchased parcels of farmland totaling 325 acres, plus an additional 205 acres of underground rights on neighboring lands, to preserve scores of sinkholes and 14 caves, at last count. He has continued to extend existing caves and explore sinkholes in search of new caves.

The Cave Farm Karst Preserve encompasses Spring Valley Caverns, a former commercial cave that offered tours only for one season before closing in 1968. Now extended to 5.5 miles, Spring Valley Caverns is the largest cave on the property, beginning in the smooth Dubuque limestone layer and extending down into rougher

The entrance building to Spring Valley Caverns at the Cave Farm Karst Preserve blends into the Fillmore County landscape.

Stewartville limestone. Several others are a half-mile long. Many have stalactites, stalagmites, flowstone, and other formations. The Sentinel, the largest and oldest cave column in Minnesota, stands in Spring Valley Caverns. Many of the caves feature streams at their lowest levels.

More than two-thirds of the cave passages are large enough to stand or stoop in. Some soar 100 feet high. Most passages are 50 to 65 feet beneath the surface, though some are as deep as 135 feet.

Ackerman's goal is to preserve these caves in their undeveloped state and make them available to cavers for education and exploration. He has no plans to install lighting or conduct commercial tours. After his death, the preserve will become a nonprofit organization.

Establishing the Cave Farm was no small feat. Ackerman purchased the land and negotiated for additional underground rights at a time when the city of Rochester was expanding and many residents were seeking rural land within easy driving distance. Land prices were escalating.

Land conservation was another issue. Ackerman began a tree-planting campaign and cleaned out the sinkholes. Some had been used as farm dumps since the nineteenth century. Today, people understand that water still seeps through southeast Minnesota's sinkholes, carrying contaminants and trash underground.

Ackerman constructed safe, secure entrances with interior emergency release features on the caves. He replaced the dilapidated building that had served as an entrance for the commercial venture at Spring Valley Caverns with a solid concrete building, dug into the hillside and faced with limestone. This building contains two entrances into the caverns. One is via an opening in the floor that allows cavers to practice their rope skills while descending more than 30 feet into the cave below. The building also contains storage space for safety and rescue equipment.

Cave exploration was costly. When there was no alternative, Ackerman used explosives judiciously to open safe entrances into caves that otherwise could be entered only by dangerous crawls—or not at all. This involved obtaining a quarry license and explaining to neighbors and officials why he needed one.

Cave exploration could also be dangerous. When digging out a sinkhole, Ackerman learned how to shore up the opening, so that the sides did not cave in. To explore one opening at the bottom of a sink, Ackerman had to wedge his way through a tight horizontal passage, filled with quicksand except for three inches at the top. The risky effort paid off: he discovered a 135-foot deep cave system, encompassing three rock layers. Construction of a safe entrance, however, was needed before most other cavers could view the discovery.

Ackerman has worked with members of the Minnesota

Speleological Survey to provide safe cave access to Boy Scouts, Girl Scouts, and other groups. MSS guides, trained in safe caving and knowledgeable about the caves' formations and features, lead trips that are either all walking tours or a combination of walking and crawling. Tours typically include information on bat protection, environmental protection, and soil and water conservation.

Directions: Directions provided at the time of appointment. Contact the Minnesota Speleological Society.

Seasons/hours: Year-round, by appointment.

Length: The Karst Preserve contains 7.5 miles of explored passages.

Precautions: The caves contain no electric lighting or concrete walkways. Wear coveralls or a sweatshirt and plan to get muddy. Bring a change of clothing and three light sources. Temperature is 48 degrees Fahrenheit.

Amenities: There are no restrooms or other amenities. The entrance building provides shelter and a place to change clothes before and after cave trips.

Contact: Minnesota Speleological Survey.
E-mail: mss@intothevoid.com.
Web site: www.intothevoid.com.

Goethite Wildlife Management Area

Near Forestville State Park, the Goethite Wildlife Management Area preserves a section of Fillmore County as habitat for beaver and other animals. Unlike most of the county, this section has numerous ponds. While ponds and lakes are often found over sinkholes in some karst regions, this is not the case in southeast Minnesota. How, then, were these ponds created?

A clue can be found in the term goethite, an iron-bearing mineral named after Johann Wolfgang von Goethe. Best known for penning Faust, the German poet and playwright also studied geology and mineralogy. The wildlife management area that bears his name once was home to many iron pit mines, though most have been closed for more than 50 years.

You can also see an iron strip-mining site on the Spring Valley–Ostrander Trail, approximately 5 miles south of Spring Valley and 1.5 miles north of the Root River County Club.

6
Masonic Park
Fillmore County

ASK AROUND IN Spring Valley about the existence of a cave in a local park, and you may get blank looks. Many residents who do know of the cave at Masonic Park have never looked for it, being warned off as children by parents alarmed by the threat of rattlesnakes. While rattlesnakes do inhabit many of the state's bluffs

The entrance to Masonic Cave is in the side of a bluff approximately 10 feet above the ground.

and crevices, cavers more often stumble on other creatures—feral cats, raccoons, other species of snakes, and even skunks. It's always a good idea to keep an eye out for such nonhuman visitors to Minnesota's bluffs and forests.

Masonic Park lies at the foot of a bluff. A creek runs along the base of the bluff, separating the park from the cliff. To reach the cave, park near the shelter and walk back on the road a short distance, recrossing the bridge over the creek. Look for a path on your right near the S-curve sign. Follow this trail toward the bluff. (Look for jack-in-the-pulpit in late May and early June and be wary of nettles in late summer.) When you reach the face of the cliff, look up. The wide cave entrance is clearly visible a dozen or so feet off the ground.

One approach is from the right along narrow "steps" and a three-inch ledge. The entrance room is about 10 square feet. At the rear is a belly-crawl passage only for professional cavers. If you attempt to climb to this cave, do not do so alone and be sure to tell a third party where you're headed.

Directions: From Rochester, follow U.S. Highway 63 south to State Highway 16. Follow 16 to Spring Valley. In town turn onto County Highway I and follow it north 1 mile to County Highway 38 (gravel). Turn right (east) and follow 38 around several 90-degree curves (which track section lines) for 3 miles to the park, on the left.

Season/hours: Year-round; weather may limit accessibility in the winter.

Length: Cave entrance room measures about ten square feet.

Precautions: The cave entrance is accessible via a 12-foot rock climb.

Amenities: Picnic shelter, tables.

Contact: Fillmore County, P.O. Box 466, 101 Fillmore Street, Preston, MN 55965.
Web site: www.co.fillmore.mn.us/,
or for park regulations, visit
www.co.fillmore.mn.us/Park%20Ordinance.htm.

7

Bear Cave Park
Olmsted County

RESEMBLING A ROCK SHELTER, the cave at Bear Cave Park near Stewartville cuts into the limestone bank of the Root River, which forms the park's southern border. The drive into the park passes several grassy playing fields before reaching two parking areas. Park in the lot on the left (east). Beyond the parking areas, roughly parallel to the county highway, you will see the edge of a wood. Bordering the wood is a line of picnic shelters and bathrooms. Beyond the picnic shelter on the far left (east), look for a paved path heading into the woods and the river beyond. Follow this trail to a bench at the top of the steep riverbank. Almost directly beneath the bench is the small, eroded cavity known as Bear Cave.

Directions: From Rochester, follow U.S. Highway 63 south to Stewartville. Turn right on County Highway 35 and watch for the park on your left.

Season/hours: Year-round.

Length: This small shelter could hold a family of bears but is too open to the elements to have ever served as a bear den.

Precautions: This small cave may be invisible in spring and at other times of high water.

Amenities: Softball fields, disk golf, hiking, picnicking.

Contact: Stewartville Parks and Recreation Department, phone: (507) 533-4360. Web site: www.stewartville.com/parks.htm.

8

Quarry Hill Park
Olmsted County

VESTIGES OF ANOTHER way of life linger along the trails and are half-hidden in the woods of Quarry Hill Park in Rochester. The city purchased most of this 270-acre parcel in 1965 from the state, which had operated the Rochester State Hospital on this and adjoining acreage since the 1870s. Hospital workers and residents—who included the indigent, chemically dependent, and mentally ill—farmed 1,000 acres, growing food for themselves and their livestock. Workers and residents also excavated a variety of underground spaces for different purposes.

The mortuary cave at Quarry Hill Park has been partially filled in by debris.

Thomas Coyne led other residents in hollowing out a Y-shaped cave in St. Peter sandstone as a food storage area in 1882. They cut about 20 bins along the sides of the 200-foot tunnel, as well as two larger rooms for storing apples and butter. One year, the bins held 9,300 bushels of potatoes, 417 bushels of rutabagas, 400 bushels each of carrots and beets, and more than 17 tons of cabbages. The

Y-shaped tunnel was more than eight feet wide and just as tall—large enough for workers to drive a wagon inside to unload the produce.

Even after canning replaced cold-storage preservation, the hospital continued to use the cave. For example, in 1941, workers stored 20 barrels of dill pickles and 15 tons of sauerkraut in the cave.

Monthly cave tours provide access to this cave, where you can see part of a poem carved into the soft stone by Thomas Coyne. Though it's partly disintegrated, you can make out several words of "Lone Mary."

Coyne also dug the smaller mortuary cave, on the east side of the cemetery, which contains more than 2,000 graves. When hospital residents died during the winter, the bodies were stored in the cave until the ground thawed and graves could be dug in the spring.

Northeast of the mortuary cave, look for the newer of the hospital's two limestone quarries, which was worked from the 1890s to 1950. One trail leads around the rim of this quarry, where you can see cracks in the rock face and occasionally the fossils of cephalopods. On the quarry floor you can see the compressor platform, dynamite house where explosives were stored, and a cap house.

To reach the older quarry, which was operated until the 1880s, follow the northwest hiking trail through the woods. South of the quarry are two kilns, as well as the remains of a cistern, and the site of a slaughterhouse and a soap house.

Workers and residents produced crushed rock from the quarried stone for the hospital and local contractors, who used it in road and building construction. The stone was transported to the crusher site in narrow-gauge railroad cars pulled by horses. The hospital operated two crushers and a pulverizer, which provided lime dust that farmers spread on their fields to neutralize acidic soil.

Only a few generations ago, quarries, kilns, and man-made caves played a major role in the life and work of many people. Today's brightly lit basement rec rooms and tiled highway tunnels seem far distant from the rough and rugged underground spaces our grandparents used.

Directions: On Silver Creek Road, off County Highway 22 (East Circle Drive) near Fourth Street South.

Season/hours: Monday through Saturday: 9 a.m. to 5 p.m.; Sunday, noon to 5 p.m.

Length: The hike from the Nature Center to the storage and mortuary caves, and quarry overlook is less than a half mile. Allow at least an additional 45 minutes to explore the old quarry.

Precautions: Trails are relatively easy, with no climbing required.

Amenities: Environmental classes, picnic shelter, playground, nature center, library, fishing, biking, baseball diamond, tennis courts, cross-country ski and snowshoe rental.

Contact: Quarry Hill Park, 701 Silver Creek Road NE, Rochester, MN 55906. Phone: (507) 281-6114. Web site: www.rochester.k12.mn.us/quarryhill/home.htm.

9 Whitewater State Park

Winona County

GEOLOGY IS THE keyword at Whitewater State Park. The structure of the earth's crust and various rock layers are evident in the park's caves, on its bluff tops, and along its trails and streams. The oldest exposed rock here dates to more than 500 million years ago—not so very long ago to geologists, who date the earth at about 4.5 to 4.6 billion years.

About 545 million years ago, most of southeastern Minnesota, southwestern Wisconsin, and Iowa was a shallow sea, where sediments built up and were cemented together over eons to form different layers of rock, such as sandstone and limestone. Generally near river mouths where sand was washed into the sea, sand layers formed, eventually becoming sandstone. Farther from river mouths or in deeper water, clay settled out, eventually becoming shale. Most distant from river mouths, sand, and mud, limestone formed from the remains of marine plants and animals. Today, fossil hunters look for indications of these ancient creatures in limestone deposits.

During the more recent Ice Age, which ended a scant 10,000 years ago, glacial meltwater eroded away the top layers of soil and rock and carved into these older layers, leaving the bluffs you see in the park today. If you walk to the top of Coyote Point, you will be standing on the Shakopee Formation. Named for Shakopee, Minnesota, where this layer is well exposed, the Shakopee is mostly dolostone (limestone in which magnesium ions have replaced some of the calcium ions).

Beneath the Shakopee, you can see Oneota dolostone as you hike up and climb down from the Coyote Point summit. Oneota is mostly

The entrance to Coyote Point Cave at Whitewater State Park is high up the bluff that forms Coyote Point.

dolostone, with less sand and shale than the Shakopee. This layer often contains fossils and caves, like Coyote Point Cave, clearly visible (though inaccessible) from the trail to the point. Nearby, several smaller cavities poke into the bluff.

Across the Whitewater River valley, another Oneota pinnacle points skyward. You can also climb to Chimney Rock and crawl into the eroded cave in its base. Chimney Rock Cave features four windows—two overlooking the valley—and is large enough for an adult to sit up. (Children enjoy the fortlike appearance of the cave, but the sharp drop-offs beneath the two windows make adult supervision a must.) From Chimney Rock Cave, you can see Coyote Point, and some geologists believe that long before the glacial meltwater cut through this valley, Chimney Rock Cave and Coyote Point Cave were part of the same large cave.

If you hike down to the Whitewater River from Chimney Rock, you will see Jordan Sandstone along the bank. The rust-colored sandstone bears the carvings of early tourists, though today the Department of Natural Resources discourages defacing the natural rock face.

Whitewater State Park contains one other significant cave; however, Trout Run Creek Cave is inaccessible since it requires rock-climbing equipment to reach and rock climbing is not permitted in the park. Still, there are other intriguing geologic features to search for in addition to Chimney Rock Cave. They include smaller cavities, a few springs, and a place where a stream disappears into the earth.

Directions: From Rochester, follow U.S. Highway 14 east (or from Winona follow 14 west) to State Highway 74. Turn north on 74 and drive about 5 miles to the park.

Season/hours: Year-round.

Length: Trail lengths vary. The Chimney Rock Trail is less than .7 mile in length and of moderate difficulty, with some steps. It typically requires one hour to complete.

Precautions: Several trails, including the Coyote Point Trail, are difficult, involving a climb to the bluff top.

Amenities: Whitewater State Park offers a year-round visitor center, rustic cabins, campgrounds, trout fishing, swimming beach, picnic areas, and recreational trails.

Contact: Whitewater State Park, Route 1, Box 256, Altura, MN 55910. Phone: (507) 932-3007. Web site: www.dnr.state.mn.us/parks_and_recreation/state_parks/whitewater/.

10
Skunk Hollow Cave
Winona County

SEVERAL MILES from Whitewater State Park lies Skunk Hollow Cave. During the summer season, a park naturalist periodically offers a caving slide show and a diagram of Skunk Hollow Cave, followed by a tour of the cave, which is on private land.

The cave is in Oneota dolostone and contains more than 100 feet of surveyed passages. It offers a good introduction to wild caving, with relatively easy crawling passages, plenty of Minnesota mud, and the opportunity to see an occasional bat.

You enter through a four-foot opening in a sinkhole in a pasture. Just northwest of this entrance is another, tinier one, used by small animals but generally useless to adult humans. Once inside, you can stand up in a six-foot dome and see roots above your head. The entrance dome is one of three small areas high enough for most adults to stand. Most passages are crawlways, some with sloping floors. At one point, a ledge partially blocks the continuation of a passage. If you are touring the cave with children, they may enjoy exploring a low-roofed chamber—sometimes termed "the Orphanage"—that's clearly off-limits to their parents.

Reservations for the tour are required. Anyone can attend the slide show, but reservations for the cave tour are limited. The cave is simply too small to permit large groups.

Directions: From Rochester follow U.S. Highway 14 east (or from Winona follow 14 west) to State Highway 74. Turn north on 74 and go about 5 miles to Whitewater State Park.

Season/hours: June through August.

Length: The cave contains more than 100 feet of passages.

Precautions: This tour involves some crawling through muddy passages. Wear pants and a long-sleeved shirt that you don't mind getting dirty. Bring a flashlight.

Amenities: There are no restrooms or other amenities.

Contact: Whitewater State Park, Route 1, Box 256, Altura, MN 55910. Phone: (507) 932-3007. Web site: www.dnr.state.mn.us/parks_and_recreation/state_parks/whitewater/.

11
Winona Bluffs
Winona County

DRIVING ON U.S. Highway 14/61 south of Winona, you can spot several caves high on the sides of the southwestern bluffs. The bluffs directly across 14/61 from the city also contain caves, though they're harder to pinpoint. Many are inaccessible.

Huff Street, which crosses 14/61 and runs between downtown Winona and the bluffs, provides the common entry point for these caves. Most are small, eroded along joints in the sandstone. Several sandstone caves can be found in the bluffs above Lower Garvin Heights Park and Holzinger Lodge, though they are easiest to reach with the assistance of an experienced guide.

This eight-foot-high cave entrance leads to a tunnel more than 100 feet long in a bluff near Winona.

One is above a trail that begins near the intersection of Huff Street and Lake Boulevard (paralleling 14/61). (This is not a cave to explore on your own or with children in tow, and remember to tell someone where you're going.) Leave your vehicle in the small lot at this intersection and walk across Lake Boulevard to tiny Lower Garvin

Heights Park. As you climb a short staircase to enter the park, you will see a fire circle on the left and a pump on the right. Just to the left of the pump, through the brush at the edge of the park, there is an abandoned stairway that climbs up the bluff. This stairway marks the beginning of a very difficult trail to a small overhang and a cave.

At the top of the stairs, climb the faint trail. Continue to follow this trail across a road and up the bluff. When the trail fades, continue climbing the steep and difficult ascent, initially through woods, then up a slippery sand and rock slope. You may see soft-drink cans, beer cans, and other debris to point you toward the cave. Eventually you will see a small sandstone overhang. As you face the bluff, the cave is above and to the left of the overhang. The entrance is narrow, but you can easily walk into the opening in the reddish sandstone. The entrance passage is approximately eight feet high, and you can walk back several yards before the passage narrows. Stop to enjoy the view of Winona before beginning the precipitous descent.

For a difficult but somewhat less treacherous climb, drive northwest on Lake Boulevard, past the old Woodlawn Cemetery and Holzinger Lodge, to Bluffside Park. Stop near the northwest edge of the park, and climb the short stairway in the stone wall to the lawn. Cross the lawn to the woods and find the path to the right of a fire circle, which ascends the bluff. Several paths crisscross this wooded slope; you are aiming for the bare sandstone face high up on the bluff. There is a small cave on this north face, visible above you as you climb. This hike is not a trek you want to try with a baby in a backpack, but it's manageable for most active adults.

After you've found the cave, you might want to try a more leisurely walk. Three primary paths begin at a trailhead just northwest of Holzinger Lodge and circle the top of the bluff, offering different levels of difficulty. Mountain bikers especially enjoy these routes, which are interconnected by a maze of unofficial side trails. The views from the top, which towers 575 feet above Winona, are magnificent, especially on a sunny fall day. If you plan to hike to the top, allow at least two hours for the round trip.

Directions: From U.S. Highway 14/61, turn south onto Huff Street. Drive to the stop sign at Lake Boulevard. Turn right on Lake Boulevard to reach the trail to the small cave on the north face of the bluff.

Season/hours: Spring to late fall, dawn to dusk. (Trails can be treacherous during winter.)

Length: Varies. Allow at least a half-hour to reach the small cave above the north end of the park.

Precautions: Some trails are very steep and challenging, unsuitable for young children and anyone with any mobility problems. Beware of slippery trails in wet conditions. Wear gloves and make sure your hands are free to grasp trees and rocks along the way.

Amenities: Playground, recreational trails.

Contact: Winona Convention and Visitors Bureau, 67 Main Street, Winona, MN 55987. Phone: (507) 452-2272. Web site: http://visitwinona.com.

12 Garvin Heights
Winona County

EASILY ACCESSIBLE BY CAR, Garvin Heights is a popular destination for visitors to Winona and offers a fantastic view of the area. It also harbors a "secret" alcove unknown to most park visitors.

Park in the lot at the top of the bluff and walk to the stone-walled observation area. Pause to take a deep breath and admire the vistas, then turn and walk down the trail to the left of the observation platform.

The cave replica at the Winona County Historical Society Museum contains fossils and petroglyphs, including this thunderbird.

Soon you will see a rock mound extending from the bluff wall. On the right side of this mound, at a right angle to the wall beneath the observation area, is a small cave. This eroded cavity takes only a minute or two to explore, though it's large enough to provide shelter in a storm and fuel dreams of long-lost lifestyles.

Directions: From U.S. Highway 14/61, turn south onto Huff Street. Drive to the stop sign at Lake Boulevard. Turn left to drive to Garvin Heights.

Season/hours: Spring to late fall, dawn to dusk.

Length: The small cave is about one tenth mile beneath the observation station.

Precautions: This short trail is somewhat steep. Wear good walking shoes.

Amenities: There are no restrooms or other amenities.

Contact: Winona Convention and Visitors Bureau, 67 Main Street, Winona, MN 55987. Phone: (507) 452-2272. Web site: http://visitwinona.com.

Winona County Historical Society Museum

The fortresslike appearance of the Winona County Historical Society Museum—it was constructed as an armory in 1915—features a sloping ramp, small windows, and tiny ornamental turrets. This square block of a building houses the county's largest array of historical artifacts, dating from before the area's early logging days to the Cold War.

Built for children, the museum's cave replica—complete with fossils and petroglyphs (early carvings)—will fascinate adults, too. Several other exhibits target children, including a teepee and a steamboat pilothouse, complete with bell.

Many exhibits are scattered about the large open room on the museum's ground floor. An encircling balcony gallery, formerly an exercise track, features a county timeline, beginning with the area's geologic history and including a history of Winona's sewers and other unlikely topics. Other highlights: a Native Americana collection, a stained glass window exhibit, and a Victorian Main Street.

The museum is downtown at 160 Johnson Street, between Third and Fourth Streets. Johnson lies northwest of and parallel to Main Street (State Highway 43). It's open year-round, Monday through Friday, 9 a.m. to 5 p.m.; Saturday and Sunday, noon to 4 p.m.

Contact the Winona County Historical Society, 160 Johnson Street, Winona, MN 55987. Phone: (507) 454-2723. Web site: www.winona.msus.edu/historical-society/.

There is a shallow cave beneath the observation area at Garvin Heights in Winona.

13

LaMoille Cave
Winona County

TODAY, LAMOILLE (rhymes with "oil") Cave is easy to miss—a two-foot-high slit near the base of a sandstone bluff bordering Trout Creek, south of the tiny community of La Moille. But the low opening was much higher a century or so ago and marked the entrance of a cave containing petroglyphs. An 1888 traveler reported carvings of fish, rattlesnakes, and thunderbirds. Since then, the floor of the cave has filled with sediment and the petroglyphs have slowly peeled away.

LaMoille Cave has a confusing history. Not only did it diminish in size over the years, there was also a rock shelter containing pottery just a quarter of a mile downstream. When construction on U.S. Highway 14/61 obliterated the rock shelter site, it was widely reported that the road construction had destroyed the cave. Not true, according to cave historian and geologist Greg Brick, who recently rediscovered the cave after consulting a stack of historic references. Besides the low "front" entrance, the cave features a small back opening and a spring in the rear.

Walking southeast from the cave along the bluff, you can also find a shallow, hollowed-out area in the sandstone face. Though it's probably too small to have interested early Native Americans as a rock shelter, the four-foot-deep cavity shows evidence of more recent occupation. Follow the trail of beer and soft-drink cans back toward the parking area.

Directions: From Winona, drive southeast on U.S. Highway 14/61 for 10 miles to the town of LaMoille. About 1 mile south of town, the highway crosses Trout Creek. A gravel drive and parking area south of the creek on the west side of the highway provides access primarily for fishermen. Follow the creek upstream about a quarter of a mile until the creek reaches the base of a low bluff and turns northerly, running along the bottom of the bluff. Look for the low cave entrance above the water line.

Season/hours: Dependent on daylight, snow accumulation, and other weather conditions.

Several rock alcoves exist in the sandstone bluff southeast of LaMoille Cave.

Length: The two-foot-high slit is long enough for an adult to lie down in.

Precautions: The short walk to the cave is through a marsh. Wear waterproof boots.

Amenities: There are no restrooms or other amenities.

Contact: Richard J. Dorer Hardwood State Forest, 1801 South Oak Street, Lake City MN 55041. Phone: (651) 345-3216. Web site: www.dnr.state.mn.us/state_ forests/sft00033/index.html.

Brownsville Caves

In the nineteenth century, residents of Mississippi River towns occasionally excavated cavities in the river bluffs as cold-storage warehouses. A row of these cliffside caves is visible at the base of the bluff facing State Highway 26 near Brownsville, south of Winona. Over the years these caves held trappers' pelts, farmers' apples, and other products. One reportedly features a shaft up to the top of the bluff. They are on private land and therefore off limits to cave explorers.

A number of manmade storage caves can be seen in the Mississippi River bluffs along State Highway 26 north of Brownsville.

14

Pipestone National Monument
Pipestone County

LONG AGO, legend has it that the Great Spirit took the form of a large bird and called together all the Native American tribes at a wall of rock. He broke off a piece of the red rock, formed it into a pipe, and smoked it. The Great Spirit told the tribes that this stone was their flesh and that they must use it only for sacred objects, according to an account recorded by artist George Catlin in 1836.

Even today, Pipestone National Monument is a spiritual experience as much as a cultural or geologic site. While anyone can walk through the historic quarry, only Native Americans can obtain a permit to quarry and sell the soft, red pipestone (also called Catlinite after the artist who brought the stone to the attention of nineteenth century scientists). Walking the Circle Trail through the quarry, you will see ancient red walls, evidence of recent digging, and occasionally small packets of bright offerings tied to overhanging limbs.

Native American tribes from as far away as the American Southwest and Canada come to excavate the pipestone in the summer and fall, when water does not collect in the ditches they dig by hand. The quarriers shovel away any soil and then break up the 10-to-15-foot layer of hard quartzite that lies over the softer pipestone. Both the quartzite and the pipestone formed 1.6 billion years ago. Quartzite is the result of great pressure and heat applied to sand, while pipestone formed from ancient clay.

The pipestone layer is typically only 14 to 18 inches thick, with only a few inches suitable for carving pipes or other artwork. The quarriers break the pipestone along natural fissures in the rock and take it out in sheets from 1 to 3 inches thick. They then cut the sheets into blocks appropriate for carving. Pipestone is about as hard as your fingernail and can be easily carved using a steel knife blade. Only the pipe bowl is crafted from pipestone; the stem is made from ash or other hardwoods.

Walking along the Circle Trail, you pass several active and inactive pits, a high wall of roughly eroded red quartzite, Pipestone Creek, and Winnewissa Falls, which rush over the red wall. You also travel briefly underground beneath a red rock archway. Markers note these natural and cultural phenomena:

- Wind and water erosion has created the Old Stone Face formation in the wall near the falls.

The red quartzite outcrops at Pipestone National Monument hide thin layers of pipestone, which can be carved into ceremonial pipes and jewelry.

- Leaping Rock was the site where young warriors would prove their valor by leaping into a chasm and placing an arrow in a crevice.
- One rock face bears the carved initials of members of the 1838 expedition to the Upper Mississippi region led by Joseph N. Nicollet.
- Old tribal shamans believed they heard voices issuing from the stone lips of the Oracle formation.
- A remnant of a tallgrass prairie changes color with the seasons as different grass and flower species bloom.

Jeffers Petroglyphs

More than 2,000 images are carved into an outcrop of Sioux quartzite about 65 miles east of Pipestone National Monument. The engravings depict humans, animals, and tools, as well as undecipherable shapes. Scientists date these carvings from 3000 B.C. to A.D. 1750.

To reach the Jeffers Petroglyphs, follow U.S. Highway 71 to County Highway 10. Follow 10 east for 3 miles to County Highway 2. Turn right on 2 and drive 1 mile to the visitor center.

Years ago, tribes came together to mine the sacred pipestone in peaceful accord. They still come in an annual pilgrimage to follow an ancient custom of careful craft and art. Today, that craft and art is open for all to witness and appreciate.

Directions: Follow I-90 west to U.S. Highway 75 at Luverne. Drive north for 25 miles to Pipestone. Signs will lead you to the monument just north of town.

Season/hours: Open daily, except Christmas Day and New Year's Day. From Memorial Day to Labor Day, the visitor center and Upper Midwest Indian Cultural Center is open from 8 a.m. to 6 p.m. Monday through Thursday and 8 a.m. to 8 p.m. Friday through Sunday. The remainder of the year, hours are 8 a.m. to 5 p.m. daily.

Length: The paved Circle Trail through the quarry is about 1 mile in length. Allow 45 minutes to take in the quarry and the marked points of interest.

Precautions: Although the monument is open year-round, freezing temperatures and snow accumulation make trekking through the quarry difficult in winter. Taking rock samples is prohibited.

Amenities: The visitor center, restrooms, and paved trail to the quarry are wheelchair accessible. A wheelchair is available at the visitor center.

Contact: Pipestone National Monument, 36 Reservation Avenue, Pipestone, MN 56164-1269. Phone: (507) 825-5464; e-mail: PIPE_Superintendent@nps.gov. Web site: www.nps.gov/pipe/index.htm.

Stone Creek

If you want to show the family what life was like on the Minnesota prairie in the nineteenth century, take them to Stone Creek. More than 130 years old, this farm homestead features three replica pioneer homes—a dugout burrowed into the landscape, a sod house, and a log cabin. A mock well and a gazebo complete the historic setting. To interest the children and entertain the adults, a tunnel, two feet in diameter, connects these structures. The park also offers a three-hole golf course, driving range, and nature trail.

From the Twin Cities, follow State Highway 7 west to Montevideo. Then follow 7 and U.S. Highway 59 northwest to State Highway 40. Turn right on 40 and drive 6.5 miles to find Stone Creek on your right. Stone Creek is open spring through fall, 2 p.m. to dusk Monday through Thursday and 10 a.m. to dusk Friday through Sunday.

Contact: Stone Creek, 9017 State Highway 40, Montevideo, MN 56265. Phone: (877) 285-6202. Web site: www.stonecreekmn.com.

15

Blue Mounds State Park

Rock County

A CURVING CLIFF OF pink and purple Sioux quartzite that forms the backbone of Blue Mounds State Park contains talus and fissures that, according to legend, once harbored the James gang. You can explore these deep fissures along the east side of the cliff on your own or on tours conducted by the park naturalist.

The quartzite was created more than one and a half billion years ago on the bottom of an ancient sea. First, sand was deposited on the sea floor and with the weight of accumulation and water turned into sandstone. Then, time, heat, and chemical reactions converted the sandstone into quartzite. Much later, glaciers gouged into the quartzite when loose rocks were dragged across the bedrock. You can see evidence of this geological history, as well as remnants of quarrying, along the park cliff.

A deep fissure cuts through the quartzite cliff at Blue Mounds State Park.

Directions: From Luverne, go north 4 miles on U.S. Highway 75. Turn east on County Highway 20 and go 1 mile to the park entrance, which leads to the main office, where you can pick up a map. To reach the talus caves, return to Highway 75 and drive south. Turn left (east) on County Highway 8 and drive to a parking area on the left.

Seasons/hours: Open year-round; however, some facilities are closed in winter.

Length: The walk from the parking area to the cliffs and back is less than 2 miles.

Precautions: The hike to the cliffs is moderately difficult, with some uphill sections.

Amenities: Camping, rock climbing, hiking, buffalo herd.

Contact: R.R. 1, Box 52, Luverne, MN 56156. Phone: (507) 283-1307. Web site:www.dnr.state.mn.us/state_parks/blue_mounds/index.html.

16

Wyalusing State Park

Grant County, Wisconsin

SITUATED ABOVE THE confluence of the Wisconsin and Mississippi Rivers, Wyalusing State Park provides a majestic panorama of the valleys carved out by these great waterways. It also offers the opportunity to safely explore one small cave and two overhangs, all from well-marked trails.

Hiking to Treasure Cave begins at Point Lookout. The paved Bluff Trail and stairways lead down the face of the cliff and through the keyhole—a narrow passage through a limestone outcrop. A short walk brings you to a sturdy ladder affixed to the wall of the cliff. When you reach the top, you can head to the right and crawl through a short passage to a rock-framed window looking out on the Wisconsin River. Or, go straight into the cave. At the back of the entrance room, a crawl tunnel leads to another room about three feet high and large enough for perhaps a dozen people to sit snugly.

Southeast of the confluence, Sand Cave—really a shelter underneath a rock overhang—is accessible via a short hike. To reach the cave by the shortest route, park at the group campsite near the naturalist's office and hike the trail about one half-mile east.

Pictured Rock Cave is a 40-foot-wide cavity, cut underneath a small waterfall feeding into a stream. To get there, take the Sugar Maple Nature Trail off of the Homestead Picnic Area and watch for the short side trail that leads to the cave. The distance from the picnic area parking lot to Pictured Rock Cave is less than one half mile.

There are other small crevices and caves along Flint Ledge Trail, a lower bluff trail that meanders along the side of the limestone cliff beneath the Bluff Trail. This challenging trail takes its name from the veins of flint (chert) visible in the rock face.

Directions: Follow U.S. 52 southeast from Rochester to U.S. 18. Follow 18 east through Prairie du Chien, Wisconsin. Turn right on County Highway C and follow it to the park.

Season/hours: Year-round, however, Treasure Cave and other locations are generally inaccessible in winter.

Length: The caves are all small; the length of the trails to reach them and return varies from 1 to 2 miles.

Precautions: Trails range in difficulty; some are steep and can be slick when wet or covered with leaves or ice. Bring a flashlight to explore the back room of Treasure Cave.

Amenities: Biking, boating, camping, canoeing, fishing, hiking, picnicking, nature center, and cross-country skiing.

Contact: Wyalusing State Park, 13081 State Park Lane, Bagley, WI 53801. Phone: (608) 996-2261. Web site: www.wyalusing.org.

Section II
Central Minnesota

EXPLORING MINNESOTA'S underside often requires scraping away the stones and sediments that cover an area's bedrock, obscure a cave entrance, or block an underground passage. The stones and sediments may be the work of a glacier's advance or retreat, the result of water or wind erosion, or the result of human development—the filling in of a sinkhole or road construction.

Learning to look with the eyes of a geologist can help in understanding Minnesota's distant past, long before there were reporters to document the shifts and seasons of the Upper Mississippi Valley. When geologists stand on a bluff and look out over a wide river valley, they appreciate the sight of a heron taking flight or the sparkling light bouncing off the water. But they have also learned to take a closer look at natural phenomena and to focus on signs that tell an ancient story. They've learned to identify the smooth, small-grained St. Peter sandstone of a river bluff, the dark outline of a fossil in a road-cut, and hundreds of other signs that bear witness to prehistory.

In central Minnesota, geologists often visually peel away the layers of glacial fill to reveal the underlying nature of the bedrock beneath our feet. When they observe the confluence of the Minnesota and Mississippi Rivers in St. Paul, they see more than just the present course of the two rivers.

The river floodplains, now filled in with glacial debris, give evidence to river channels that once carried large volumes of water from melting glaciers. During the last Ice Age, the Mississippi River carved multiple channels through the Twin Cities region before settling on its present course. These debris-filled channels underlie the cities today.

The Glacial River Warren, ancestor of the Minnesota River, began as an outlet of Glacial Lake Agassiz, a huge inland lake formed by melting glaciers that once covered northwestern Minnesota and parts of North Dakota, South Dakota, Saskatchewan, Manitoba, and Ontario. After the lake emptied, the mouth of the river migrated southward. The Minnesota River today is but a shadow of its mighty ancestor. Yet the broad river valley remains to record the strength and size of Glacial River Warren.

St. Anthony Falls also documents the glacial history of the Twin Cities. More than 10,000 years ago, this waterfall was part of River Warren Falls, a huge cataract downstream from the confluence. Slowly the falls eroded its riverbed and moved upstream. When it reached the confluence of the two channels at Fort Snelling, it divided into two falls, one retreating upstream on the Glacial River Warren and the other up the ancient Mississippi River. Today, St. Anthony Falls remains fixed, captured by modern erosion control, about eight miles above Fort Snelling. The Glacial River Warren Falls was extinguished when the river intersected another channel (now buried) about two miles from the fort.

The rivers cut through bedrock consisting of more than 100 feet of Ordovician St. Peter sandstone capped by a thinner layer of Platteville limestone. The St. Peter is relatively soft, and area rivers and streams have carved deep gorges through the fine-grained white rock. Passages and caves have been eroded beneath the cities and along the riverbanks, for example, along the banks of the Mississippi River upstream and downstream from Fort Snelling.

City engineers and developers have excavated tunnels in the soft sandstone for a variety of purposes. The relative ease of excavation meant that projects that might have been too costly elsewhere were feasible in the Twin Cities. For example, the University of Minnesota dug two 600-foot caverns under its West Bank Campus on the Mississippi River to house the Minnesota Library Access Center in 1999. The Platteville limestone provides a stable roof for the caverns, which hold the Elmer L. Andersen Library and other collections.

The sites listed in this section include man-made and natural features in a variety of rock layers. Many relate directly or indirectly to the area's three major rivers—the Minnesota, Mississippi, and St. Croix—and many are in Ordovician sandstones, limestones, and shales. A few sites, however, showcase older rocks, such as the 1.6-billion-year-old granite of Cold Spring Quarry and the volcanic basalt of Interstate State Park.

All these sites invite a closer look and challenge us to focus on the long view so as to decipher ancient stories. ■

17
Carlson Lime Kiln
Goodhue County

THE VIEW IS WORTH the climb to the top of Barn Bluff, more than 340 feet above the city of Red Wing. One of the most impressive of the bluffs along the Mississippi River, Barn Bluff is listed on the National Register of Historic Places.

The partially walled-up entrance of Carlson's Lime Kiln on Barn Bluff in Red Wing testifies to the site's history as a quarry.

Early travelers, including Jonathan Carver and Henry David Thoreau, raved about the panorama spread below and beyond this vantage point. Artist Henry Lewis recorded an earlier Dakota legend about it. According to the story, the bluff was twice its present size when two Dakota villages argued over its possession. The Great Spirit resolved their quarrel by dividing the mountain into two halves: one village received Barn Bluff in Red Wing and the other got Sugar Loaf in Winona.

In the nineteenth century, Barn Bluff and Sugar Loaf were quarrying sites. Here, the Carlson Lime Kiln produced various products, including materials for bridge construction. You can still see the kiln, now with a partially walled-up entrance, along a trail on the top of the bluff.

Directions: From downtown Red Wing, follow East Fifth Street to the east side of Barn Bluff. Look for a stairway heading up the bluff on your left and a commemorative plaque on your right. Follow the stairway up the bluff. At the top of the stairway, the trail divides, providing two paths to the summit. Take the path to the right and continue to bear right, skirting the bluff. The kiln will be on your left.

Season/hours: Year-round, weather permitting.

Length: A hike to the summit and back takes about an hour.

Precautions: Steep stairway and moderately difficult trail.

Amenities: The city of Red Wing is known for its pottery, antiques, and parks.

Contact: Red Wing Visitors and Convention Bureau, 418 Levee Street, Red Wing, MN 55066. Phone: (651) 385-5934. Web site: www.redwing.org/.

18

Red Wing Memorial Park
Goodhue County

THE NARROW SNAKE of a road leading through dark woods to the top of Sorin's Bluff opens suddenly into a broad-mouthed prairie, covered with grasses and wildflowers waving a welcome. The road circles this plateau, and in a quarter-mile arrives at an observation area, with a historical plaque and stone barrier at the edge of the cliff. This prominence offers an aerial view of downtown Red Wing's streets and buildings.

An unusual tunnel-shaped kiln is located off a trail on the other side of the bluff top. Follow the one-way circle road to the short drive on your right leading to a picnic area. Enter the large parking area with the iron railing at the end. Stand with your back against the iron railing, facing the parking area. The trailhead is just ahead on your

This entrance leads to a tunnel-shaped kiln in Red Wing's Memorial Park.

left. The recreational trail slopes downward into an old limestone quarry. After approximately a quarter mile, you will see a quarry wall on your right, with a side trail that curves along the base of this wall and then rejoins the main trail. At the point where the siding-style trail rejoins the main trail, look to your left to find a faint trail sloping downward. Follow it, but beware of the steep drop-off on your right. About 75 yards beneath the main trail, look for the opening of a man-made tunnel on your left.

This four-foot opening leads to a kiln, though it differs from many historic kilns in that it is actually a horseshoe-shaped tunnel. You can find the other, much smaller opening farther along the trail at the base of a tree. If you look down from this point in leaf-off conditions, you will see Wilkinson Street below at the base of the bluff. When men worked in the quarrying operations here long ago, the view would have been wilder and the noises louder. Today, standing quietly on this spot, you are likely to hear only an occasional voice as mountain bikers ride by on the trail above.

In another section of the park, a sandstone cave begins as a pasageway and then opens into a room. The cave ceiling lies in an Ordovician layer where you can see fossilized wave ripple marks. This small cave is difficult to reach. Attempt it only with an experienced guide, and tell someone where you've gone and when you expect to return.

Directions: From U.S. Highway 61 in downtown Red Wing, turn south on State Highway 58. Turn left on East Seventh Street, following the signs to Memorial Park. Turn right into park and drive up the switchback road to top of bluff.

Season/hours: Spring through late fall, dawn to dusk.

Length: The man-made tunnel is perhaps 20 feet long.

Precautions: The trail to the natural cave is hazardous, particularly in wet conditions.

Amenities: The city of Red Wing is known for its pottery, antiques, and parks.

Contact: Red Wing Visitors and Convention Bureau, 418 Levee Street, Red Wing, MN 55066. Phone: (651) 385-5934. Web site: www.redwing.org/.

Carver's Cave

A remnant of Minnesota's most historically significant cave now lies undisturbed on private land at the base of Dayton's Bluff below Indian Mounds Park. To reach a plaque marking the importance of Carver's Cave, park in the lot between Cherry and Plum Streets and walk to the observation area.

Named for Captain Jonathan Carver, an explorer who "discovered" the cave in 1766, it once contained petroglyphs—possibly left by the Hopewellian culture, thought to have created the mounds on the top of the bluff around A.D. 400. When Carver arrived, the then resident Dakota (Sioux) regarded the cave as a dwelling of spirits.

Carver partially explored the cave and recorded his findings. He went through an opening ten feet across and three feet high to enter a room 30 feet wide with a ceiling 20 feet high at its highest point. Near the entrance Carver found petroglyphs, some covered over with moss. About 60 feet from the entrance he came to a lake and did not venture into the farther reaches of the cave.

Years later, early nineteenth-century explorers had more difficulty in finding and entering Carver's Cave, due to fallen rock and debris. In the 1850s efforts were made to excavate the opening in the talus—or sloping pile of rocks—and remove sand from the cave, widening and deepening it. In 1867, the Minnesota Historical Society held a Carver Centennial Celebration at the site, which had by then become a popular tourist destination.

But about a year later, the Chicago and St. Paul Railroad destroyed part of the entrance, laying additional tracks between the cave and the Mississippi River. In the years that followed, fallen rocks and debris again threatened the cave, although Boy Scouts and other groups periodically located and explored it.

As a U.S. Bicentennial project, city and state officials searched for, found, and reopened the cave; a group of Native Americans entered it for the first time in many years. To protect the historic place, steel doors were subsequently fixed over the entrance. Since then, the doors have all but disappeared behind mounting sand and debris. Once again, Carver's Cave awaits its next official rediscovery.

But cavers haven't forgotten its presence. St. Paul musician John Knowles has even written and produced a series of "cave music" CDs, inspired by Carver's Cave.

In 2002, the city purchased a parcel including Carver's Cave from the Burlington, Northern and Santa Fe Railroad, planning to clean up and preserve the area. It is possible that future visitors may be able to hike to and see the entrance to the cave, which Native Americans view as a sacred site.

The purchased parcel also includes a natural cave used by the Joseph Schmidt Brewery to store beer in the nineteenth century.

19 Wabasha Street Caves
Ramsey County

THE CARPETED dining area, brick walls, stucco ceiling, and fireplace make St. Paul's Wabasha Street Caves among the most tastefully appointed anywhere. Oh, yes, there is also a 60-foot bar, sculpture from a mansion garden, tile floor in the cocktail area, and gleaming hardwood dance floor. Several cave rooms are even heated.

The entrance to the Wabasha Street Caves abuts this bluff in downtown St. Paul.

Like the other man-made sandstone caves in the bluff south of the Mississippi River, today's Wabasha Street Caves began as silica mines in the nineteenth century. But unlike most of the other caves—which have been used primarily for growing mushrooms, aging cheese, storing equipment, and other traditional commercial purposes—three caves in this group of seven occupy an additional notorious page in history, as a speakeasy and gangster hangout during Prohibition.

The story began benignly enough. After the caves had been carved out of the bluff and silica mining ceased, Albert Mouchenotte and his friends, Louis Lambert and Charles Etienne, immigrated to St. Paul from France and began growing mushrooms in these caves in the 1890s. A mushroom-shaped doorway between two cave rooms bears witness to this era.

In the 1920s, Mouchenotte gave the caves to his daughter Josephine, who had married William. The Lehmanns converted three of the caves into a nightclub called

Castle Royal in 1933. The club provided a venue for nationally recognized entertainers like the Dorsey Brothers of big band fame. And, it became popular with gangsters like John Dillinger, Baby Face Nelson, and Ma Barker and her boys, thanks to Police Chief John J. O'Connor, who made St. Paul a safe city for gangsters—if they did not commit crimes there.

For a while O'Connor's plan worked. Gangsters felt at home in St. Paul, where they kept speakeasies, gambling halls, and brothels in business, while committing bank robberies in other cities. But eventually, the plan backfired, when gangsters kidnapped William Hamm Jr., a brewery owner, and held him for ransom in 1933. A year earlier, two police officers had been killed when gangsters robbed the Third Northwestern Bank of Minneapolis, fleeing to St. Paul. Public and

Mushrooms, Mines, and Mayhem

The bluffs lining the Mississippi River upstream of the Wabasha Street Caves (see p. 56) in St. Paul were once pockmarked with scores of caves dug into the St. Peter sandstone. Many dated to the nineteenth century, and most were the result of mining the sand.

For example, after the Ford Motor Company's assembly plant in the Highland Park section of St. Paul was constructed in 1925, it mined the sandstone beneath the plant to use in producing glass for automobile windows. As the mining tunnels grew, tracks were laid and tram cars used to haul the sand, until the mining operations ceased in the 1950s.

Many of the older caves along the river, including the Wabasha Street Caves, were used for growing mushrooms; so the area is sometimes called "mushroom valley," even though the caves were also used for aging cheese and beer and storing a variety of goods.

You can still see openings to several sandstone caves in Lilydale Regional Park, although trespassing in them is forbidden. You can explore, however, for fossils in the park's old brickyard operations, where mining has exposed fossil beds in the Decorah shale dating back hundreds of millions of years to a period when this area was at the bottom of an ancient sea. To purchase a fossil-hunting permit, call the St. Paul Division of Parks and Recreation at (651) 632-5111.

In another cave farther west along Shepard Road, the St. Paul Jaycees have operated a Tunnel of Terror every October for twenty years. On Crosby Farm Road (the entrance road to the Crosby Farm Nature Area), the cave is across from the marina and is used as boat storage at other times of the year. The Halloween cave tours feature a variety of haunting stations along a circuitous route. If you go, be prepared for a long line—this is a popular annual event! To learn more, visit the tunnel's web site: www.tunnelofterror.org.

St. Paul's sandstone river bluffs contain many man-made caves, including several behind a parking lot on Concord Avenue near its intersection with Wabasha Street.

media opinion turned against the gangsters, and many were brought to trial.

Yet even as public outrage mounted, the nightclub continued to cater to the gangsters. The brick walls and stucco ceiling were installed in 1933, and the fireplace was added the following year. Alabaster and gold chandeliers decorated the ceiling. At some point, a tunnel was dug at a steep angle up to the top of bluff more than 60 feet above the caves. Now the tunnel is blocked at the top by the basement of a house, but perhaps it was used as a way to move liquor and other supplies in and out of the caves.

People still tell stories about the nightclub. One tale involved a young woman who stayed late to clean and close up after most customers and all the other workers had gone home for the night. Four customers were still playing cards in the fireplace room, and she cleaned the other areas while they finished their game. About 4 a.m., she heard angry voices and then gunshots. When she ran to the room, she found three men lying apparently dead on the floor with blood splattered on the white tablecloths. The young woman ran to the entrance, where there was a telephone, and called the police. She then stayed at the entrance to await their arrival. When the policemen came and walked back to the fireplace room, there were no dead bodies or blood on the tablecloths.

The most likely explanation is that the gunman carried out the bodies from another exit with the help of accomplices—or even the police themselves. One thing is certain: despite efforts to repair the damage, the fireplace bears marks that look suspiciously like bullet tracks.

Eventually, of course, the gangster era came to an end, thanks in part to public outrage and the work of J. Edgar Hoover and the FBI. As the nation prepared for World War II, the nightclub closed in 1940.

The space remained empty for a time, but after the war Land O' Lakes leased cave space for aging blue cheese. A cement floor was added in the rooms used for cheese aging. Despite many improvements over the years, a couple of the cave rooms still have a sand floor. One is filled with debris bulldozed in by the city after a Mississippi River flood.

All of the cave rooms are spacious, and usable space totals 12,000 square feet. Ceilings are 15 to 20 feet high. There's room for 250 people to dine comfortably and space for many more if the occasion is a simple reception. The space still seems admirably suited for a nightclub, and, indeed, it served as a disco during that dance craze from 1977 through 1981.

Current owners Donna and Steve Bremer purchased the caves in 1991, intending to use them as storage and office space for their construction company. But, once again, the space itself seemed to demand a different use. Today, the caves are open for twice-weekly tours, swing-dance sessions on Thursday nights, and weddings, murder mystery dinners, and other parties the rest of the week.

The nightclub spirit remains and lingers. Occasionally, a worker sees a mist that coalesces into a human form or a shape that glides through one of the deserted rooms. People report hearing music, laughter, and glasses clinking when the caves are supposed to be empty and silent. Some people believe gangsters are buried here.

Certainly, memories are buried here—the kind of memories that trigger fantasy, imagination, even nightmares.

Directions: The caves are on Wabasha Street, about two and one half blocks south of the Mississippi River and one half block south of the Wabasha Street–Plato Boulevard intersection.

Season/hours: Cave tours are offered Thursday at 5 p.m. and Saturday at 11.

Length: Allow 45 minutes for the walking tour.

Precautions: Dress casually; wear walking shoes.

Amenities: The city offers a variety of restaurants, shops, and parks.

Contact: Wabasha Street Caves, 215 Wabasha Street South, St. Paul, MN 55107-1805. Phone: (651) 292-1220. Web site: www.wabashastreetcaves.com.

Two Buried Caves: Fountain and Chute's

Before the development of the city of St. Paul, Fountain Cave opened in the side of a pristine sandstone bluff less than 400 feet from the Mississippi River. A sparkling stream issued from the cave entrance, flowing through a deep ravine to the river. Early explorers wrote about the cave's beauty and the waterfall it contained.

In 1837 voyageur Pierre "Pig's Eye" Parrant built a saloon and cabin at the mouth of the stream, selling whiskey to soldiers from Fort Snelling. But Parrant's business was short-lived, and eventually the area around the cave became part of the fort's expanded lands.

By mid-century, visitors headed to the area not to buy whiskey but for a different type of diversion. They toured the underground attraction, taking advantage of vendors selling ice cream and lights to illumine their path through Fountain Cave. "It was opened as a show cave back in 1852, complete with ticket booth," noted cave historian Greg Brick. This date marks the first commercial cave operation in Minnesota.

Slowly and inexorably, the city developed, the flow of water through the 1,150-foot cave subsided, and debris accumulated, closing the entrance. When Shepard Road was constructed in the early 1960s, workers filled in the ravine, obliterating any remaining sign of the entrance to Fountain Cave. A plaque on the recreational trail bordering Shepard Road, just southwest of Randolph Avenue, commemorates the role of Fountain Cave in St. Paul history.

Several other sandstone caves lay hidden beneath the Twin Cities. The most famous, Carver's Cave is mentioned elsewhere (p. 55). Another worthy of mention is Chute's Cave in St. Paul, also a former commercial cave. Hollowed out of sandstone with a limestone ceiling, Chute's Cave contained flowstone and other formations that once attracted late nineteenth-century tourists.

A decade before boat tours began in 1876, Chute's Cave was the site of a well-publicized hoax. A "Reuben Nesmith" claimed to have found "Nesmith's Cave" while digging a potato bin. He reportedly uncovered an iron trap door opening on a spiral staircase that led into a passage and cave encompassing several rooms. The story, which reached the pages of the New York Herald, listed such finds as a human skeleton, an altar, and copper implements. Eventually, the story died, and the connection between Nesmith's Cave and Chute's Cave was lost among the tall tales of the cities' history.

Today, Chute's Cave lies largely forgotten, hidden by roads, buildings, sewer tunnels, and other development that have sprouted in the area.

20

Fort Snelling State Park
Hennepin County

"HISTORY UNDER the Floor Boards," a multimedia exhibit, provides an archaeologist's view of the excavation in the "basement" of the historic Fort Snelling Officers' Quarters. The exhibit pinpoints the bases of two limestone fireplaces, walls, and other features revealing how the building evolved since its original construction in the 1820s. An 11-minute video explains how archaeologists dated the addition of the second fireplace from the 1840s and analyzed more than 28,000 fragments of dishes, utensils, marbles, and other artifacts to piece together a picture of life in the fort. Young would-be archaeologists are usually intrigued by several vignettes that depict real-life exchanges among family members living in these quarters in the 1820s. (Oops! There goes another marble, fork, or hairpin—lost through a crack in the floorboards.)

Children appreciate the many interpretative programs at the fort, especially the soldiers firing muskets and cannons, blacksmith shop, and fur trade encampment. A small regiment of interpretative guides in period costumes takes visitors back to life on the Minnesota frontier. Guided tours have sometimes included Quarry Island (also known as Wakan or High Rock Island), to see clues left by a nineteenth-century quarrying operation.

Adults are often more impressed by the scope of the fort's reconstruction, which began in the 1950s. At the time, only four buildings remained standing (officers' quarters, round tower, hexagonal tower, and the commandant's house). The restoration took fifteen years and $4 million.

The longtime survival of the fort depended considerably on the reliable water source of Coldwater Spring, located upstream on the west bank of the Mississippi River. U.S. soldiers from the fort took over the spring from Native Americans in 1820 and built Camp Coldwater. In the 1880s, they constructed a limestone well tower, pump house, and reservoir. The spring supplied water to the fort until World War II.

For centuries before the soldiers arrived, the spring supplied inspiration as well as water to Native Americans. The Dakota viewed Coldwater Spring as the dwelling place of the god of the underworld.

The largest spring in the Twin Cities, Coldwater is at least 10,000 years old.

Directions: Located in Minneapolis at State Highway 5 and State Highway 55, above the confluence of the Minnesota and Mississippi Rivers.

Season/hours: Memorial Day through Labor Day, Wednesday through Saturday, 10 a.m. to 5 p.m. and Sunday, noon to 5 p.m.; open weekends in May, September, and October, Saturday, 10 a.m. to 5 p.m., and Sunday, noon to 5 p.m. Closed November through April.

Length: Allow several hours to explore the historic fort.

Precautions: The journey from the parking area to the fort is an easy 10-minute walk.

Amenities: The Fort comprises more than a dozen restored buildings, including a hospital, officers' quarters, barracks, schoolhouse, blacksmith and harness shops. The History Center features exhibits and films, and hiking trails run along the Mississippi River beneath the Fort.

Contact: Fort Snelling History Center, St. Paul, MN 55111. Phone: (612) 726-1171. Web site: www.mnhs.org/places/sites/hfs/index.html.

21
Joseph Wolf Caves
Washington County

IF YOU'VE NOT RECENTLY toured the caves in the bluff in downtown Stillwater, you may want to revisit them. If you've never walked through these unique underground rooms, now is a good time to plan a weekend trip to the riverside town.

Once an integral part of the Joseph Wolf Brewery, the caves and the adjoining buildings have been renovated by 215 Wabasha Properties of St. Paul, the same company that owns and operates the Wabasha Street Caves tours (p. 56). This is good news for visitors, since the new owners have repaired the caves' water-damaged museum, researched the history of the caves to provide accurate tour information, and remodeled the complex. The stucco has been removed from the exte-

The Joseph Wolf Caves are a prominent feature of downtown Stillwater.

rior to reveal original sandstone walls, and a branch of Grumpy Steve's Coffee Shop has been transported from the St. Paul location.

The caves date to the early days of the Wolf Brewery, begun by Martin Wolf, who immigrated to Stillwater from Switzerland in 1868. He located his brewery at a spring near the edge of the St. Croix River in the thriving lumber town. Wolf constructed several frame buildings including the brewery and a boarding house for lumber workers. Over the years, the company went through several forms:

- In 1871, Martin Wolf turned the buildings and the brewery operations over to his brother, Joseph, and Joseph Tanner, who renamed the brewery Wolf Tanner and Company. A year later, the wooden buildings were destroyed by fire. Following the blaze, Wolf and Tanner constructed the stone buildings that stand today.
- In 1876, Joseph Wolf bought out Tanner and renamed the company Empire Brewery.
- In 1896 Wolf brought his three sons and two sons-in-law into the business, simultaneously changing the name to the Joseph Wolf Company.

By 1887, the brewery was producing 6,000 barrels of beer annually. The beer was stored and aged in enlarged natural limestone caves beneath the brewery.

The three cave rooms near the entrance of the underground complex were originally cellars dug into the bluff adjacent to the brewery. A row of three homes built on concrete slabs ("Slab Alley") had their "basements" in the bluff behind the houses. For readers who enjoy interesting trivia, Bing Crosby's mother once lived in one of these houses.

With Prohibition, the Wolf Brewery ceased beer production in 1920, converting to soda production. The soda operation failed, and after his death, Joseph Wolf's children shut down the company in 1925.

Twenty years later, Thomas Curtis bought the property. An implement dealer, Curtis first moved a portion of his business to the new location. Soon, however, he had a better idea. Stillwater was becoming a tourist destination, so Curtis connected the three cellars to the brewery caves, flooded them, and offered boat rides through the caves. He also stocked the ponds with trout and allowed children to fish in the caves. The kids could fish for free, but they had to pay a dollar for every trout they caught. They could also pay a small fee to feed the fish.

In the 1950s the caves were designated a fallout shelter, and school children would run to them during drills. A fireplace in the caves may date from this period.

Curtis owned the caves until 1971. Tours (sans boats) have been offered regularly through the years. Most recently, the old brewery building was home to an Italian restaurant. The new owners plan for a smaller restaurant in this space, along with retail shops.

While the brewery buildings now look more than ever as they did in their boom times in the late nineteenth and early twentieth centuries, the area around the brewery has changed dramatically. The row houses disappeared in 1933. The St. Croix River has shifted eastward, as fill has been added along its Stillwater bank. And Main Street was widened in the 1980s.

But the spring still runs into a pool in the caves. This stream then leaves the cave pool, running under more cave rooms and resurfacing outside near the cave entrance, before dipping beneath the street and flowing to the river. The caves once more appear as they did in their brewery heyday—full of artifacts, stories, and memories of a rough-and-ready era.

Directions: From the Twin Cities, follow I-35 east to State Highway 95 north. Stillwater is on 95, about 10 miles north of I-35. Follow 95 downtown to the old Wolf Brewery at the intersection of South Main Street and Nelson Street.

Season/hours: April through November, 10 a.m. to 6 p.m., daily, excluding Tuesday. Limited winter hours.

Length: Allow 35 minutes for the quarter-mile tour.

Precautions: Dress casually; wear walking shoes.

Amenities: The city offers a variety of restaurants, shops, and parks.

Contact: Joseph Wolf Caves, 402 S. Main Street, Stillwater, MN 55082. Phone: (651) 292-1220. Web site: www.wabashastreetcaves.com/cavetour.html.

22

Interstate State Park

Chisago County

AT FIRST GLANCE, the pothole-studded landscape of Interstate State Park seems to belong to some other state—or planet. But then a glance down into the gorge of the St. Croix River reminds you that this, too, is Minnesota.

The cliffs above the river are basalt, resulting from lava flows dating back 1.1 billion years, following a line of volcanic eruptions from Taylors Falls to Lake Superior. Then, from about 500 billion years ago to 70 million years ago, advancing and retreating seas washed through this area, leaving behind sedimentary rocks, fossils, and ripple marked rocks that are still visible today.

A mere 10,000 years ago, as the last glacier melted and retreated north, rushing water carved the St. Croix Valley. Pounding whirlpools blasted sand and silt against the bedrock, eroding the huge potholes on the bluff above the present-day St. Croix River. Interstate State Park contains more potholes than anywhere else in the world, as well as the deepest explored pothole. The Bottomless Pit is more than 60 feet deep, and some of the unexplored potholes in the park may be even deeper. The Minnesota Department of Natural Resources pumps water from some of the potholes to better view their depth.

You can walk among these giant cauldrons on the self-guided Pothole Trail; or tour them in summer with the park naturalist, who points out evidence of two different glaciers, ten lava flows, and other geologic indications.

A highlight of the trail is a short tunnel that leads into the Bake

Boom Hollow Cave

North of Stillwater in the low bluff along the St. Croix River, early settlers dug a storage cave into the Franconia sandstone. It lies just upstream from a spring outlet into the river and extends more than 60 feet into the bluff.

The cave's name dates from the lumber era, when lumber companies installed a boom, or barrier, across this section of the river to corral the logs, which were processed at mills downstream.

The Bake Oven was formed when two whirlpools merged along the St. Croix River in Interstate State Park.

Oven, a formation created by two whirlpools. Their 35-foot shafts descend at an angle, somewhat like the two sides of a misshapen V. The entrance to this formation resembles a baker's oven door. Inside, the eroded walls are smooth and the interior cool.

When you complete the Pothole Trail, there is probably plenty of time to explore the woodlands and prairies, which provide shelter for a surprising number of animals and plants, including native cactus and several endangered species. As you walk on the cliffs above the St. Croix, watch kayakers shoot the rapids and envision the logjams that choked the river in the late nineteenth century. The park was established in 1895 in a joint effort with Wisconsin and was the nation's first interstate park. A park sticker from either state permits access to the park on both sides of the river.

Directions: From the Twin Cities, follow I-35 north to U.S. 8. Then take 8 northeast to the park.

Season/hours: Open daily; however, naturalist programs and "pothole tours are available only during the summer.

Length: he stone surfaced Pothole Trail is a relatively easy quarter-mile walk, although it includes several stone stairways.

Precautions: Although campsites are available year-round, campers must walk or ski to sites after measurable snow has fallen.

Amenities: Picnic shelters, volleyball, camping, hiking, canoeing, fishing.

Contact: Interstate State Park, P.O. Box 254, Taylors Falls, MN 55084. Phone: (651) 465-5711. Web site: www.dnr.state.mn.us/parks_and_recreation/state_parks/interstate/index.html.

Stearns History Museum

"On Solid Ground," the quarry replica and exhibit at Stearns History Museum in St. Cloud, is for people who like to know how things work. The exhibit explains the beginning of quarrying in the area in the 1870s and depicts the history of the industry through 1960.

Before 1900, quarrymen primarily depended on hand tools and dynamite to create dimension stone for building. After the turn of the century, improved drills and bits enable them to more efficiently extract rectangular blocks from a quarry. Also about this time concrete began to replace dimension stone in building, although thin decorative facings of stone continued to be used. Soon electric charges and other power tools further facilitated the work of the quarrymen. Derricks enabled them to lift the large blocks, called **loaves,** from the quarry and place them onto trains. Today, when a loaf reaches the plant, it is **slabbed,** using wire saws and lasers.

During the early twentieth century, a quarry typically employed four categories of workers: stone cutters, polishers, blacksmiths (to make and maintain tools), and **lumpers,** or unskilled laborers.

In the heyday of quarrying in the early twentieth century as many as 25 companies operated quarries in this region. Today, few remain. But St. Cloud is home to the largest granite company in the world; the Cold Spring Granite Company operates 18 quarries in the area.

Granite comes in a variety of colors—primarily gray, pink, and red—and is usually composed of feldspar, quartz, and other minerals. This hard, igneous rock is coarse-grained and often light in color. You can see it used on many public and corporate buildings today in almost any city.

And you can still see the remnants of 25 historic quarries in a county park in suburban St. Cloud: Quarry Park and Nature Preserve, 1802 County Road 137, Waite Park, (320) 255-6172, contains 130 acres and offers swimming in a former quarry, trout fishing, scenic overlooks, hiking, biking, and cross-country skiing trails.

To reach the museum from the Twin Cities, follow I-94 north to State Highway 15 at St. Cloud. Follow 15 north to State Highway 23 east. Turn right on 23 and follow it to 33rd Avenue. Turn left on 33rd and follow it to the museum. Open Monday through Saturday, 10 a.m. to 4 p.m.; Sunday, noon to 4 p.m.; closed on major holidays.

Contact Stearns History Museum, 235 S. 33rd Avenue, St. Cloud, MN 56301-3752, (320) 253-8424. E-mail: info@stearns-museum.org. Web site: www.stearns-museum.org.

23 Crystal Cave
Pierce County, Wisconsin

THE LONGEST CAVE in Wisconsin has close connections to Minnesota. Located 45 miles east of St. Paul, Crystal Cave draws tens of thousands of visitors each year, many from Minnesota. The Minnesota Speleological Survey has volunteered hundreds of hours to clean formations and clear out new passages. The latter is especially slow work, involving digging by hand and carrying dirt out in buckets. A Minnesota Department of Natural Resources small mammal specialist is studying the cave's bats. During the winter months, researchers band the bats' wings, weigh them, and survey their locations in the cave.

Crystal Cave is home to all four of the cave bat species common to Minnesota and Wisconsin. More than two thirds (70 percent) of Crystal Cave's bats are **little brown myotis;** almost one third (30 percent) are **eastern pipistrelles.** The cave also harbors small numbers of big brown bats and **northern myotis.** "If you don't like bats, come in July and August," suggested Crystal Cave co-owner Blaze Cunningham. "If you want to see them, come in spring and fall."

While fascinating, the diversity of its bat population is not the most significant feature of Crystal Cave. What stands out are the highly decorated walls, sculpted several hundred thousand years ago. Still, this is a young cave, according to Cunningham. One formation has been dated at about 10,500 years old—definitely young by cave standards.

Visitors can approach the stalactites and stalagmites for close-up viewing, though mesh prevents curious fingers from touching the formations. Oils on human hands can interrupt the growth of fragile speleothems.

Tours begin with the recent history of the cave. Story has it that Crystal Cave was discovered in 1881 by a teenage boy chasing a squirrel near his farm home. When the squirrel disappeared down a sinkhole, the boy probed the hole with a stick. When he dropped the stick and it disappeared in the depths, the boy, William R. Vanasse, went home to tell his brother, George.

Next day, the two boys brought a rope and a lantern to the spot and lowered themselves down a "slide" into a clay- and debris-filled room. From there they dropped into the main room on the second of the cave's three levels. In several directions, the boys saw shallow en-

A relatively young cave, Crystal Cave still contains stalactites, stalagmites, flowstone, and other formations.

trances to clay-filled galleries on the upper level.

For decades, Sander's Corner Cave, as it was then called, remained partially filled. Adventurers occasionally visited and explored, but no one realized the size and potential of the Vanasse find—until Henry A. Friede came on the scene in 1941.

An advertising agency manager and amateur geologist from Eau Claire, Friede had been studying caves in the area in hopes of finding one equal to that of Wisconsin's Cave of the Mounds, discovered in 1939. He decided that of the several possible sites in the region, the most potential lay in the development of Sander's Corner Cave. In November 1941, a crew began removing the silt and debris that filled the cave, for the most part completing the job by April of the following year.

Work began on an entrance building, but even before it had been completed, Friede opened Crystal Cave to the public in June of 1942. Some 4,000 visitors showed up on opening weekend. The new name referred to the quartz crystals found throughout the cave. Since then, more than one million visitors have toured Crystal Cave, which has been further cleaned and explored.

The cave has changed ownership four times. Current owners Blaze Cunningham and Jean Place are geologists who have emphasized the educational aspects of the cave.

Crystal Cave is formed in dolostone and contains iron ore, flint, and, of course, quartz crystals. There are also two snail-shaped fossils about the size of a nickel. Both of these gastropods date from the time the area was part of a giant inland sea.

Ramps make the cave walk easy and safe. One of the first things visitors see after descending a stairway to the first level of the cave is a series of man-made pools, with water cascading from one to the next. The pools attractively solve a perennial problem for commercial cave operators: how to collect and divert water running through the cave, in order to keep trails dry—or at least to keep them free of flowing water. In a living cave, water still enters along cracks and crevices, continuing to build stalactites, stalagmites, flowstone, and other formations. Generally these cracks run northeast/southwest with many cross joints.

Caverns to explore include:

- The Ballroom is the largest chamber, where many bats hang out in winter, and where more than one couple has exchanged marriage vows.
- The Spook Room features unusual "faces" in the walls and is often the kids' favorite.
- Most kids also enjoy the Story Room, where guides weave a spell by telling such tall tales as that of Cave Man Charley, who once lived in the cave and participated in boat races on an underground river. In the dim depths, with the fantastical formations of Charley and the remainders of his boat clearly visible, even adults almost believe in the myth. Underground, it certainly seems more plausible than, say, a blue ox.
- The walls of the Wish Room are filled with illite. According to Jean Place, illite is a general term for mica-like clay minerals often found with shales. Clays containing illite are used extensively for makeup, since they absorb facial oils; in construction to stop leaks in dams, for example; in china; and in medicines. Visitors are invited to place a coin in the soft illite walls and make a wish. The longer their coin sticks to the wall, the more likely their wish will come true.

Crystal Cave is a popular destination for Minnesota school groups. The guides—most are trained college students—can lead groups of up to 29 people each, and groups can leave as frequently as every 10 minutes. Six groups can be in the cave at any one time, so the cave easily accommodates busloads of visitors. The guides try to respond to the interests of each group, spending extra attention on fossils or rocks or bats or other topics.

Visitors get even more individualized attention on Adventure Tours, conducted from November through March. During this two-hour trip through a less developed area of the cave you experience real caving—crawling through narrow, twisting passages and into large, domed chambers, eventually popping out through the bats' doorway.

This tour begins with a caving orientation session, which requires belly crawling through a box to ensure that you will be able to navigate the tightest passages. The registration fee includes hard hat and headlight.

The Adventure Tour is not for anyone taller than 6 feet, 4 inches, or heavier than 220 pounds. Minimum age is 14, and children 14 to 18 must be accompanied by an adult

Directions: From St. Paul, go east on I-94 to Exit 19, turn south on U.S. Highway 63 and go 7 miles. Follow the signs toward Spring Valley. Crystal Cave is about 1 mile west of Spring Valley on State Highway 29.

Season/hours: Open April 1 through May 15, 9 a.m. to 5 p.m. weekends; May 15 through Labor Day, 9 a.m. to 6 p.m. daily; September through October, 9 a.m. to 5 p.m. daily. (Last tour leaves 30 minutes before closing.) Fee.

Length: Tour lasts about one hour. Current, official length of the cave is 2,065 feet, with approximately another 1,000 feet of additional passageways.

Precautions: Light jacket is recommended; temperature is 50 degrees year-round.

Amenities: Picnic tables, restrooms, and gift shop.

Contact: Crystal Cave, W2465 Highway 29, Spring Valley, WI 54767. Phone: (800) 236-CAVE or (715) 778-4414. E-mail: cavebats@svtel.net. Web site: www.cavern.com/crystalcave.

Section III Northern Minnesota

"GOING UNDERGROUND" HAS A different meaning in northern Minnesota than in the southeast region of the state. In the North, underground attractions are usually mines, not caves. They are generally in the northeast quadrant of the state and are large, requiring neither crawling nor stooping. The mine sites listed in this section are also more accessible than many of the caves described earlier; there are fewer challenging hikes, steep stairways, and rocky climbs.

In the northeast in particular, there is no karst, no limestone or dolostone. When you drive northeast in Minnesota, mountains replace level or gently undulating land and forests overwhelm fields.

Several glaciers advanced, retreated, and advanced again over northeast Minnesota, scraping and scouring the land. There is little glacial fill here, since most glacial debris was deposited along the edges of the thick ice sheets, particularly in their terminal moraines. When they melted, the glaciers left exposed much Precambrian rock.

These rocks include granites formed deep within the earth's crust, basalts created during ancient lava flows, and sandstones born in prehistoric shallow seas. They also include the iron formations of Minnesota's Cuyuna, Mesabi, and Vermilion Ranges, stretching in a diagonal swath across northeast Minnesota from Crosby to Ely.

While each range has played an important role in the state's mining history, their ore formations are not identical. In the Vermilion Range, for example, the ore bodies tend to extend vertically, like upright bean pods buried underground. In contrast, the ore bodies in the Mesabi Range tend to lie within 500 feet of the surface, spread laterally over a wide area.

Iron is found in a variety of rocks within these formations. Much of the highest-grade, black-red **hematite**, consisting of 65 percent iron, has been mined; however, technology advancements have made mining the less rich **taconite** profitable. Taconite contains magnetite as well as hematite, and modern production methods separate the iron magnetically. Even though iron is the fourth most common element of the earth's crust (after oxygen, silicon, and aluminum), it is usually not found in the high concentrations present in Minnesota's iron ranges.

Ore was first shipped from the Vermilion Range in 1884, though it had been discovered years earlier during the gold explorations of the 1860s. In their quest for gold, prospectors initially ignored the mineral that would prove more valuable to Minnesota over the long term. Less than a decade after the first shipment of iron ore left the Vermilion Range for Two Harbors, the first iron ore shipment left the Mesabi Range; the first iron ore was shipped from the Cuyuna Range in 1911.

The majority of the sites in this section commemorate the state's iron mining and processing, although one listing marks the Rainy Lake Gold Rush along the Minnesota-Canada border and others document man-made and natural sites along the North Shore, around Duluth, and farther south at Banning State Park.

Yet the iron mines, especially the pit mines of the Mesabi Range, have had the most widespread impact on the landscape. Surface mining created many gargantuan pits. Some of these have been reclaimed as lakes and are today popular among scuba divers and fishers. Surface mining also created huge tailings piles, which in many places have been planted with trees and now appear as young forests. Not since the glaciers crunched across these lands has the topography been so altered.

24
Banning State Park
Pine County

BANNING STATE PARK provides an unusual diversity of underground and in-ground features to explore. There are potholes and an historic quarry, complete with partial buildings and quarrying sites. The Log Creek Arches were formed from eroded kettles. Robinson's Ice Cave is closed to humans to protect this bat hibernaculum. And a small, sandstone cave breathes out cool air on a July day.

You can hike through the old quarry along the self-guided Banning Quarry Trail, stopping along the way to view Dragon's Tooth Rapids of the Kettle River and the ruins of the quarry's cutting shed, powerhouse, and artesian well. The well provided income from bottled water sold in Chicago and elsewhere. Between 1892 and 1912 these woodlands reverberated with the sounds of blasting and cutting, as well as the thundering of the Short Line of the St. Paul–Duluth Railroad. Hundreds of quarrymen worked at drilling, moving, and shaping the pink sandstone into building stones, paving stones, curbs, and bridge abutments. The quarrying boom ended as the supply of good stone diminished, along with the demand for sandstone as a building material. Increasingly, structural steel and concrete replaced the more natural sandstone.

North of the old quarry, a trail along the east bank of the Kettle River provides access to the Log Creek Arches. There also are potholes in the bottom of the Kettle River. They occasionally pull canoes and canoeists into the centrifugal currents. In addition to Log Creek and the Kettle River, the park encompasses Wolf Creek, six streams, and three springs.

Directions: From the Twin Cities, follow I-35 north to Exit 195. Follow the signs to the park entrance.

Season/hours: Year-round, however, many campsites, water, and restrooms are closed during winter.

Length: The Banning Quarry Trail is a moderately easy 1.8-mile walk through woodlands and rock formations along the Kettle River.

Precautions: Trails vary in difficulty. The Banning Quarry Trail is moderately easy, encompassing small hills and ravines.

Amenities: Banning State Park offers a camper cabin,

campgrounds and watercraft campsites, boat landings, picnic areas, overlooks, and recreational trails.

Contact: Banning State Park, P.O. Box 643, Sandstone, MN 55072. Phone: (320) 245-2668. Web site: www.dnr.state.mn.us/parks_and_recreation/state_parks/banning/.

In winter, large ice stalagmites often form in Robinson's Ice Cave. Photo courtesy of Minnesota Department of Natural Resources.

25

Robinson's Ice Cave
Pine County

GATED TO PROTECT the bats that live there, Robinson's Ice Cave is not open to the public; however, you can approach the entrance and sometimes see bats leaving the cave at dusk in the summer. You can also visit a smaller cave en route to Robinson's.

Although it's part of Banning State Park, Robinson's Ice Cave and another smaller sandstone cave are most easily reached from Robinson Park on the east edge of Sandstone. From the parking lot, walk north along the Kettle River trail. Walk underneath the railroad trestle to an area where the trail appears to end in a tumble of sandstone boulders. In the bluff on your left, look for the opening to a small cave that extends perhaps 20 feet into the hillside. Ahead, the trail continues on the other side of the tumbled rocks. Cross the boardwalk and two small ravines. Then, look for a small, unmarked trail on the left that leads to the bluff and Robinson's Ice Cave.

This cave is intentionally a challenge to locate. It is accessible primarily to researchers who study the three species of bats that hibernate there: Keen's myotis (a somewhat rare species), little brown bat, and big brown bat. Robinson's Ice Cave is about 200 feet deep and in some places 20 feet in height.

Why is it called "Ice Cave"? In winter, cold air is trapped in the cave's wet rooms, forming large ice stalagmites.

Directions: From the Twin Cities, follow I-35 north to Exit 195. Robinson's Ice Cave and the small cave are accessible from a Sandstone city park connected to the state park. Located on the west bank of the Kettle River, this park is on Old Wagon Road, off State Highway 123 at the eastern edge of town.

Season/hours: Inaccessible during the winter.

Length: The hike to Robinson's Ice Cave is about 1.5 miles.

Precautions: The trail to the cave is rugged and unmarked, not recommended for young children or people with limited mobility.

Amenities: Picnic area and recreational trails.

Contact: Banning State Park, P.O. Box 643, Sandstone, MN 55072. Phone: (320) 245-2668. Web site: www.dnr.state.mn.us/parks_and_recreation/state_parks/banning/.

26
Croft Mine Historical Park
Crow Wing County

MINNESOTA'S SOMETIMES forgotten Cuyuna Range lies southwest of the better-known Mesabi and Vermilion Ranges, hidden beneath tracts of lakes and woodlands. Yet its story is worth remembering and the Croft Mine Park is worth a stop the next time you are heading to the Brainerd Lakes area.

Situated on the original mine site, the park contains a museum, mining equipment, miner's cottage, and a tour through a mine replica that's as real as advice from retired miners can make it. The simulation, finished in 1988, provides a first-hand experience of the sometimes dangerous business of mining underground on the Cuyuna. The only thing you don't do is pick up a hammer or drill or heft the ore into a waiting tramcar.

Mining began later here than in the Mesabi and Vermilion Ranges. Cuyler Adams organized the Orelands Mining Company in 1903 with William McGonagle and William C. White. Cuyler's dog was named Una—hence the Cuyuna town name. The first mine in this range was the Kennedy Mine near the town of Cuyuna. Soon, other mines opened in the area.

Croft means "farm" in Swedish, and true to its name, this mine was discovered on a Swedish farm. The Croft Mine was built in 1912, began shipping ore in 1916, and operated until 1934. Its original shaft was 634 feet deep. During its heyday in 1920, the Croft Mine employed 165 men.

Today, you can wander about the 17-acre site to view large mining equipment, a dynamite "house" resembling a short silo, and a honeymoon row cottage built by George H. Crosby, founder of the town. The white cottage—the last one not remodeled and still remaining of a several-block section of row houses—was moved to the park when it was developed. The only original building still standing in the park is the dry house, where the miners changed their clothes before and after their shifts underground, and which today houses the park's museum.

Guided tours begin at the ticket booth and wind aboveground on a path that lies directly above one of the underground tunnels. The tunnels, along with the mineshafts, are now flooded. Walking along the mowed path, you can see the sites where the hoist house and shaft house once stood, narrow gauge tracks where ore trams once ran, and

The Croft Mine replica features a real tramcar—and tracks.

a side trail leading to a beaver pond. The pond was created when a sinkhole developed following dynamiting in the mine.

Alongside the ramp leading into the reconstructed shaft house, you can spot ore buckets and a 2.5-ton **skip car** used to transport the mined ore. In the shaft house, a modern elevator waits to take you "underground." Originally, steam engines powered the elevator and air and water pumping systems.

When the elevator comes to a stop, the doors open into a realistic replica of a **drift**, or horizontal tunnel. The drift owes its authentic appearance to carefully crafted cement and fiberglass "rock" walls of the same dimensions as the real thing. Still, the tunnel's comforts would have surprised the miners who once worked 10-hour shifts in the Croft. The cement floor is smooth underfoot, water does not drip from walls and puddle on the floor, and electric lights provide plenty of illumination. The temperature is 60 degrees Fahrenheit, a full 10 degrees warmer than when the real mine was in operation.

The Croft, like other underground mines, was operated all year. Ore mined during the winter months, however, was stockpiled for shipment in warmer weather.

Cuyuna Range miners had to dig through more than 100 feet of overburden to get to the iron ore. They would sink a shaft into the adjacent slate and other more stable rock, and then dig drifts to the softer ore bodies.

Coleraine Historic Walking Tour

Pick up a free brochure describing a walking tour of Historic Coleraine at the city hall, which is next to the high school, or at the Depot Visitor Center in nearby Grand Rapids. A booklet providing more detailed descriptions of each property on the tour is also for sale at the city hall or Coleraine Public Library.

Named for Thomas Cole, president of U.S. Steel's Oliver Mining Company, Coleraine differed from the typical mining town in several respects—thanks largely to the efforts of John C. Greenway.

U.S. Steel hired Greenway, a friend of President Theodore Roosevelt and a former member of his Rough Riders, to expand mining operations in the area and develop a company town. But Greenway's idea of a company town was far more progressive than most. He replaced tract housing, where rows of identical homes lined up like soldiers, with homes as individual as the people who eventually inhabited them. Greenway also insisted that housing applicants be screened before being allowed to move in. Streets were named after his friends or other area developers.

Coleraine's walking tour passes a company housing district, as well as the larger homes of the general mining superintendent, the area's first doctor, and the first school superintendent. Also on the route: a log church, the Carnegie library, a log lodge, and the Coleraine ski jump site. Greenway and U.S. Steel built the wooden jump for the town in 1909-10.

Coleraine is about 300 miles north of the Twin Cities. You can take U.S. Highway 169 all the way; the town is a few miles northeast of Grand Rapids.

To view a more typical mining town, you can visit Bovey, right next to Coleraine along U.S. Highway 169. Drive through its streets to find several rows of miners' homes, all identical and closely crowded. Some of these were boarding houses, with different establishments catering to workers of different nationalities. According to a Bovey resident, one bed was sometimes rented to three different men, who slept in shifts.

Contact the Grand Rapids Area Chamber of Commerce, One NW Third St., Grand Rapids, MN 55744. Phone: (800) 472-6366. Web site: www.grandmn.com/.

At Croft, miners constructed two shafts and six drifts. They used one shaft to bring workers and ore in and out of the mine; they carried timbers into the mine through the second shaft. As the miners dug into the soft ore, they used timbers to shore up the cavities they formed.

Digging upward, the miners created diagonal passages called manrises into subdrifts where much of the work was done. The **manrise** contained a **manway**, providing access to the miners and an ore chute.

The ore in the Croft Mine had a 55-60 percent iron content, the highest on the Cuyuna Range. The ore was moved by mine car underground, then hoisted to the surface and loaded into narrow-gauge rail cars for transport to a commercial rail line.

Also visible in the replica drift is a first aid room furnished with medicines, bandages, and a stretcher. Emergency evacuation was on a timber cart through the timber shaft, with the cart knocking against the walls of the shaft on the slow, jerky ride to daylight.

The miners used a variety of tools. In the early days, most work was done by hand. One miner would hold a long handle with a drill bit at the end against the rock wall, while another miner hammered the end of the long handle. The miner holding the drill would regularly turn the tool and be careful to keep his hands out of the way of the other fellow's hammer in the dim candlelight.

During World War I, improvements were introduced. For example, the miners switched to carbide lamps, which provided better light than candles. As at the Soudan Mine (see page 109), the miners used a **tugger** (winch) to power a **scraper bucket**, which dragged the ore along the passage to the chute where it would drop into a mine car for transport to the main shaft. There's even a Longyear steam-driven core drill here, used to collect samples for rock and mineral analysis.

At end of the tunnel replica, you pass an engineer's office, spread with tools and papers. The engineer determined where and how to proceed as the maze of underground passages grew.

The exit of the tunnel replica serves as one entrance to the park's museum, which showcases an array of mining tools, clothing, ore samples, and other artifacts. One of the most mesmerizing displays is that of the 1924 Milford Mine disaster. Forty-one miners were killed when a section collapsed, sending a crush of mud and water into the Milford's passages. In half an hour, the Milford was flooded. Only seven men working in the Milford's upper levels escaped as a swamp and pond caved into the mine.

Equally mesmerizing is the story of Vicenzo Rocca, a miner who quit the Milford two weeks before the collapse. Tired of working in waist-deep water at the Milford, Rocca feared a potential catastrophe and went to work at the Croft. Turned out he was right to do so. The collapse occurred just 15 minutes before quitting time at the end of the day shift on February 5.

Directions: **From the Twin Cities, follow U.S. Highway 169 north to State Highway 18 along the western shore of Mille Lacs Lake. Take 18 west to State Highway 6. Turn right (north) on 6 and drive to Crosby, through the intersection with State Highway 210 to the entrance, about 0.5 miles north of 210. The park is also accessible from Second Avenue, eight blocks north of 210.**

Season/hours: **Open from 10 a.m. to 6 p.m. daily, Memorial Day weekend through Labor Day.**

Length: Allow an hour or two to explore the park, take the tour, and see the museum exhibits.

Precautions: Wear walking shoes. The temperature in the mine replica is 60 degrees.

Amenities: Museum, library, restrooms, picnic tables, playground equipment.

Contact: Croft Mine Historical Park, c/o Crosby-Ironton Joint Powers Board, Box 97, Ironton, MN 56455. Phone: (218) 546-5466 or (218) 546-5625. Web site: www.exploreminnesota.com/. (Search under "Things to Do.")

27 Hill Annex Mine State Park

Itasca County

FROM THE MOMENT you get out of the car you realize that Hill Annex Mine differs from Minnesota's other state parks. There are no recreational trails, no map to help you hike through the park. The park office is in the miners' two-story clubhouse, built in the 1930s. The miners could join this clubhouse, where they held meetings and social activities.

Currently the clubhouse contains mining exhibits, as well as the park office, which shares the main floor with displays explaining mining operations, miners' equipment, and how mining changed over time. Upstairs, there is a historic mineral analysis lab, a geologist's office, the mine superintendent's office, a typical miner's bedroom, and even a schoolroom for the miners' children.

The only way to see the abandoned Hill Annex Mine today is by a bus tour that circles the open pit and includes several stops where you can disembark to see various overlooks, mine buildings, and equipment. All guides are knowledgeable about mine operations; many are retired miners.

First leased for mineral exploration in 1892, Hill Annex was not actively mined until 1913, when it began as an underground mine but quickly converted to strip-mining. Then it was seriously excavated, producing 63 tons of iron ore before closing in 1978. At the height of its operations Hill Annex was the sixth largest producer of iron ore in Minnesota.

Early in the tour the air-conditioned bus passes the site of the mine's "location," or company town, at one time home to more than a dozen mining families living in small houses built by the mining company. As the mine grew, the homes were moved in the 1960s to further expand the pit. Workers then had the option of buying these homes at rock-bottom prices.

When the mine closed, the pit was a half-mile wide at its widest point and had reached a depth of 500 feet. Today the bottom 200-300 feet are underwater, and the Department of Natural Resources regularly pumps it out. Groundwater seeps into the pit and a waterfall flows out of the side of the pit from an underground river. More than 1.5 million gallons a day enter the pit from this river.

The cold water in the pit is home to perch and even a few northern and walleye. Reclamation has brought wildlife back to the area, including the peregrine falcon, wolf, deer, fox, and an occasional bear.

Standing on an overlook along the edge you can feel a stiff breeze blowing from the pit. Most pit overlooks are very windy; currents rise from the pit bottom, fueled by temperature differences between the low and high elevations.

You can see almost 50 miles in several directions. Beyond the pit, the rolling landscape is shaped largely by humans; many hills are former mine dumps. Aside from the ridge marking the Continental Divide, there's little natural ground left on the Mesabi Range.

The tour showcases many remnants from the mining era. There's a 1936 Great Northern caboose and a wooden snowplow car that was also used to push brush and debris off the tracks. You can see giant shovels, trucks, and the speeder cars that carried workers in and out of the mine via rail in the 1950s and 1960s. One of the few remaining wooden water towers is also on the tour route. Here, workers filled up the watering trucks used to keep the dust down in the pit and on the roads leading from it.

Equipment breakdowns occurred routinely. Minor repairs were done as soon as possible, but major repairs were left for the winter months. Hill Annex contains a train shop, specifically for the repair of train engines and cars, and your tour bus can drive right through the large doors. The floor of the shop is laid with wooden bricks, in contrast to the cement floors found in other mine shops. If a worker dropped a gear on the softer wood floor, the iron gear teeth were less likely to break than if they hit more solid cement.

Accidents were common. Workers dropped equipment, forgot to watch for moving vehicles, and became careless around machinery. Most workers wore steel-toed boots and hardhats for minimal protection.

Workers specialized, taking on specific job responsibilities and developing particular expertise in their fields. A mid-twentieth century

The tour at Hill Annex Mine State Park offers panoramic views of a former pit mine, now partially water filled.

pit mine in the Iron Ranges employed mechanics, carpenters, truck drivers, and the highly skilled shovel operators, who typically received a higher salary. While mining was generally a male domain, in World War II fully half of Hill Mine's workers were women.

Prior to the war, each miner was issued a brass identification tag when he was hired. At the beginning of his shift, he would hand the tag to the "tool checker" and receive in exchange the tools he needed that day. At the end of the shift, he turned in his tools and retrieved his brass tag. Come payday, the miner would present his tag to receive his pay. You occasionally see these round brass tags—usually no more than two inches in diameter with a hole near the edge—in Minnesota antique shops.

The Hill Annex workers mined nonmagnetic hematite, as opposed to magnetite, which is separated from other elements magnetically and formed into taconite pellets. (See Minntac tour, page 104.) At Hill Annex, miners extracted the iron by first crushing the ore and then pouring it into a pudding that included water, steel, scrap iron, and other materials. The iron sinks and the material that floats to the surface is discarded. The process is repeated in additional slurries where the heavier iron again settles out.

Presently, Minnesota primarily provides taconite to the nation's steel mills. But as the resources of magnetic ore shrink and technology improves, the scenario may change. It's conceivable that in the distant future, industry may return to the lean ore remaining in the Hill Annex Mine and in the surrounding dumps. Stockpiled ore may be a revenue source at Hill Annex and the Mine may return to active status.

Directions: From Grand Rapids, take U.S. Highway 169 northeast for 15 miles to Calumet. Follow the signs to the park. Grand Rapids is about 175 miles north of the Twin Cities.

Season/hours: Open from Memorial Day weekend through Labor Day. Hours: The visitor center is open daily from 9:30 a.m. to 6 p.m. Mine tours are offered from 10 a.m. to 4 p.m.

Length: Tours last one and one-half hours.

Precautions: Wear walking shoes if you plan to get off the bus to explore the replica mining buildings.

Amenities: The visitor center offers exhibits, videos about mining, and restrooms. Near the Hill Annex entrance is a Calumet city park with picnic tables, a shelter, and playground equipment.

Contact: Hill Annex Mine State Park, P. O. Box 376, Calumet, MN 55716. Phone: (218) 247-7215. Web site: www.dnr.state.mn.us/parks_and_recreation/state_parks/hill_annex_mine/index.html.

28
Hull Rust Mahoning Mine Overlook
St. Louis County

THE MINE THAT grew and grew—engulfing a town in the process—first shipped ore in 1895. Miners have excavated more than 1.4 billion tons of earth from the Hull Rust Mahoning or Hull Rust Mine in more than a century of operation. Now a National Historic Landmark, the world's biggest open pit iron ore mine is more than 3 miles long and 2 miles wide at its widest point. It is 600 feet deep.

You can get a good view of this Grand Canyon of the North from the observation area north of Hibbing. The pit encompasses many historic open pit mines and many of these are now water filled. The deepest, the Scranton pit, is filled with water to a depth of 380 feet.

Sections of the Hull Rust are still being worked. You can see giant drills, dinosaur-like shovels more than 30 feet tall, and a parade of monstrous trucks valued at $2.5 million each snaking slowly in and out of the pit. They head to the Hibbing Taconite Company on the far horizon, where the ore is processed into taconite pellets for shipment to the steel mills of Indiana and elsewhere.

A large part of the Hull Rust is beneath the groundwater table. Water enters the pit at a rate of about 16,000 gallons a minute. Pumps maintain the water at the desired level, enabling mining operations to continue.

The observation area north of Hibbing includes a visitor center and a path along the edge of the pit with several observation stations. Shovels, giant trucks, and other equipment are also on display. Standing inside of one of the huge shovels or in front of a 240-ton truck helps you comprehend the magnitude of these mining operations.

On your drive to the observation area, notice the grid of laid-out streets, curbs, and foundations of Old Hibbing. As the mine grew, it threatened the town itself, and in 1918 many buildings in what was then the heart of Old Hibbing were moved two miles south, to the site of the mining town called Alice Location.

Frank Hibbing held the first lease to mine ore on the site in 1891, and by 1893 the first railroad arrived at the location. The first ore was carried over the Duluth, Mississippi River and Northern Logging Railroad to Swan River and there transferred to the Duluth and Winnipeg line, which took the ore to the docks at Superior, Wiscon-

The Hull Rust Mahoning Mine, known as the Grand Canyon of the North, encompassed a town as it grew to its current size of more than three miles long and two miles wide.

sin, for shipment to the nation's steel mills. The original Mahoning Mine was the first open pit mine on the Mesabi Range, and soon strip mining replaced underground mining everywhere on the Mesabi.

Ownership of the mine changed several times in its early years, and in 1899 James J. Hill of the Great Northern Railway bought Mahoning Mine, along with the logging railroad and a tract of land. After Andrew Carnegie and Judge Gary organized U.S. Steel in 1901, its Oliver Iron Mining division took over operations at the Hull Rust pit, which is still controlled by U.S. Steel.

It has taken more than a century to dig this massive pit, which grew as the United States developed into a world leader economically and politically. This multilobed pit—and the miners who worked it—helped the nation develop and win two world wars. During peak production in the 1940s, almost one quarter of all iron ore mined in the United States came from the Hull Rust pit. Looking out over the mammoth hole, you can almost grasp its importance to Minnesota and the nation.

Directions: From the Twin Cities, follow I-35 north for about 135 miles to State Highway 33 near Cloquet. Take 33 north to U.S. Highway 53 and drive to the town of Virginia, a total distance of about 60 miles. In Virginia, turn west on U.S. Highway 169 and drive 26 miles to Hibbing. Then follow 169 and State Highway 73 north to Howard Street. Turn left on Howard and drive to Third Avenue East. Turn right on Third and follow the signs to Hull Rust Mine.

Season/hours: The observation area is open from 9 a.m. to 7 p.m. daily from mid-May through September.

Length: The world's largest open pit iron ore mine is more than 3 miles long, up to 2 miles wide; maximum depth is 600 feet.

Precautions: The mine overlook is accessible to all.

Amenities: Visitor center, restrooms, trail, picnic area, BMX track. Many of the skilled guides are retired miners.

Contact: Tourist Center Seniors, 1202 E. Howard Street, Hibbing, MN 55746. Phone: (218) 262-4166. Web site: www.irontrail.org/hullrust.htm.

Minnesota Museum of Mining

The most fascinating underground attractions at the Minnesota Museum of Mining in Chisholm are actually aboveground. The steam and electric shovels, iron ore train, mine pump, and mineral samples will thrill any excavator or train enthusiast, or budding geologist. Granted, there is a replica of an underground drift, but it resembles a small basement more than a mine. The multitude of informative artifacts positioned in the yard and buildings fronting the castle are far more interesting.

Castle? The stone monument, complete with turrets, dates to the Great Depression. The Works Progress Administration (WPA) project originally formed part of a larger, community athletic complex that was converted to a museum about 50 years ago. It's an appropriate venue for things historic, particularly anything related to stone, minerals, and mining.

Exhibits in the castle and in one of the two other display building provide step-by-step explanations for exploring, plumbing, drilling, blasting, ore washing, and other mining processes. A map depicts worldwide mineral deposits and another display describes ore transportation by rail and lake. There's a model of a taconite plant and a mining town, featuring mine office, blacksmith shop, newspaper office, school, and other replicas. Early firefighting equipment, a 1916 White bus (forerunner of the ones operated by Greyhound, which was founded in nearby Hibbing), early household furniture, and industrial and farm equipment fill the second display building.

Many of the outdoor displays focus on iron mining:

- A 1904 100-ton Duluth, Missabi and Iron Range locomotive "pulling" ore cars and a caboose.
- A Prescott steam-driven water pump that kept the old Monroe Mine shaft dry and supplied Chisholm with a water source.
- A 1910 Atlantic Steam Shovel, a type used to dig the Panama Canal and the only remaining example in existence.
- A 35-ton Euclid dump truck, the type often used in the pit mines during the post–World War II years.
- A 127-ton Wabco Electrohaul dump truck, now small in comparison to the trucks used in today's pit mines, but demonstrating one stage in the ever-increasing size of the massive mining trucks.

The museum is in Chisholm, a few miles northeast of Hibbing. Chisholm is about 190 miles north of the Twin Cities via I-35, Highways 33, 53, and 169. Once on State Highway 73 in Chisholm, watch for the water tower and the museum castle on your left, located in Memorial Park at the top of Main Street. The museum is open daily from 9 a.m. to 5 p.m., Memorial Day through Labor Day.

Contact the Chisholm Area Chamber of Commerce, 102nd Avenue NW, Chisholm, MN 55719. Phone: (800) 422-0806 or (218) 254-7930. Web site: www.chisholmmnchamber.com.

Helmets hanging on an underground timber appear ready to be worn when the next miners' shift begins at the Minnesota Museum of Mining in Chisholm.

29

Ironworld Discovery Center
St. Louis County

One of the highlights of a visit to the Ironworld Discovery Center is a ride aboard a 1928 trolley to the re-created Glen Location or mining camp. The trolley, plastered inside with photos of the past, leads you in mind and spirit to another era. Along the way, you pass a water-filled, 2-mile-long pit of the Glen Mine. You travel near enough to the edge of the pit to catch wonderful views of the rugged gorge.

Other sights on the journey also take you back to Minnesota's iron mining heyday. You spot an 1890 engine house replica and an actual 1890 railroad turntable. There's even a wooden water tower dating from 1910.

When your trolley pulls up at the Wilpen Depot, you step out into another time. The historic mining camp is complete with bunkhouse, bright yellow miner's home, and various mining accoutrements, as well as the Wilpen Depot. Both the depot and tiny miner's house—once home to 13 people—were moved to this site.

An interpreter welcomes you to the 1915 mining camp, introduces you to the town's characters, and spins tall tales of Minnesota's wild mining frontier. You learn that the original Glen Location once had 50 homes and that they were often built close to the edge of the pit, to shorten the miners' walk to work.

Then, you have time to stroll about the town, walk around the little yellow house, examine miners' clothing and personal items in the bunkhouse, and generally make yourself at home in the 1915 camp. You also have the unusual opportunity to hike into a mine pit by following a 1.7-mile self-guided trail offering several rest stops along the way.

When you've explored the camp and the pit to your satisfaction, simply return to the depot and wait for the return trolley. During the summer season, the trolley stops at the depot every half hour.

Back in the main Ironworld complex, you can take advantage of many different encounters with history:

- Outdoor living history exhibits include a Trapper's Cabin, an Ojibwe camp, and a Pioneer Homestead, complete with chickens.
- The Iron Range Interpretive Center features mineral exhibits, logging displays, and costumed interpreters ready to explain the area's immigration history and to entertain with the music

At Iron World Discovery Center, riders on a 1928 trolley can view this two-mile-long pit of the Glen Mine on their way to the re-created Glen Location, the camp where miners once lived.

and foods of other countries.

- A Civilian Conservation Corps History Center contains an amazing array of vehicles and other artifacts related to the work of the organization in Minnesota, which employed more than 84,000 young men in 100 camps between 1933 and 1941.

Paved walkways and a bubbling stream meander through the park.

Mineland Reclamation

Sherwood Forest Campground at Gilbert provides campers a beach, docks, boat launch, fishing piers, two pavilions, a hiking trail, and a playground. Campers can fish, swim, water ski, scuba dive, hike, and play volleyball. Each campsite affords a view of Lake Ore-Be-Gone, a 140-acre mining pit lake. The lake formed when the former Gilbert, Schley, and Pettit pit mines filled with water. The campground was developed with the assistance of the Mineland Reclamation Division of the Iron Range Resources and Rehabilitation Board, established in 1977.

The Reclamation Division also aided the development of the Keewatin Industrial Park and addressed safety issues surrounding former mines—for example, capping old shafts and fencing former pit mines. But revegetation has been perhaps its largest project. The division grows several hundred thousand tree seedlings annually for planting in former mine lands.

Mines operating today generally plant filled tailings areas and pit slopes with trees and grasses to prevent erosion and control dust. Driving through the Mesabi Iron Range, you note many forests planted within the past quarter century.

Directions: From the Twin Cities, follow I-35 north for about 135 miles to State Highway 33 near Cloquet. Take 33 north to U.S. Highway 53 and continue north to the town of Virginia, a total distance of about 60 miles. In Virginia, turn west on U.S. Highway 169 and drive west about 22 miles to Chisholm. Ironworld is on 169 in Chisholm.

Season/hours: Open daily May through September, 9:30 a.m. to 5 p.m.

Length: To be fair to yourself (and the interests of your traveling companions), allow the better part of a day to explore Ironworld.

Precautions: Wear good walking shoes to tour the outdoor exhibits and the Glen Mine location.

Amenities: Restaurant, living history exhibits, Minnesota CCC History Museum, Iron Range Research Center Library, miniature golf with a mining theme, playground equipment.

Contact: Ironworld Discovery Center, West U.S. Highway 169, P.O. Box 392, Chisholm, MN 55719. Phone: (800) 372-6437. Web site: www.ironworld.com.

30

Mountain Iron Historic Overlook

St. Louis County

A NATIONAL HISTORIC LANDMARK, the Mountain Iron Historic Overlook offers a view of the place where iron ore was found on the Mesabi Range. Captain James Nichols made the discovery while working for Leonidas Merritt and his brothers in 1890. The precise spot on the far slope of an open pit now lies on land owned by U.S. Steel's Minntac plant. After Nichols's discovery, a shaft was sunk and the first ore shipped to Duluth in 1892. In addition to the viewing station, this small park features mining equipment and a steam-powered 1910 Baldwin locomotive used to haul ore to Duluth.

Mountain Iron has established another memorial to its mining origins. A two-and-one-half-ton statue of Leonidas Merritt stands in front of the Mountain Iron Public Library. You can't miss it.

Directions: From the Twin Cities, follow I-35 north for about 135 miles to State Highway 33 near Cloquet. Take 33 north to U.S. Highway 53 and continue north to the town of Virginia, a total distance of about 60 miles. In Virginia, turn west on U.S. Highway 169 and drive 3 miles to Mountain Iron. The small park is at Mountain Avenue and Locomotive Street.

Season/hours: Open year-round.

Length: The Overlook is about one half mile from the spot where Captain Nichols discovered iron.

Precautions: The Overlook and park are completely accessible.

Amenities: The Overlook, the locomotive, and the mining equipment are the primary attractions in this tiny park.

Contact: Iron Trail Convention and Visitors Bureau, 403 N. 1st Street, Virginia, MN 55972. Phone: (800) 777-8497. Web site: www.irontrail.org/mtiron.htm.

Prairie grasses and trees now cover portions of the pits of the former Hull-Nelson Mine, visible from the Leonidas Overlook near Eveleth.

31 Leonidas Overlook
St. Louis County

IN ANY SEASON, you first notice the vibrant colors of Leonidas Overlook: the sapphire lake, vermilion cliffs, and emerald forests. The vistas sparkle dramatically in the slanting rays of a late afternoon or on any sunny day after the leaves have begun to turn.

Nature's beauty at its best—or is it? The truth is, the overlook is the highest man-made point on the Mesabi Range and the lake will be gone in a generation or two. Created from "overburden," or surface

At Mountain Iron Historic Overlook, a white slash painted on the pit wall marks the spot where Captain James Nichols discovered iron on the Mesabi Range while working for Leonidas Merritt and his brothers in 1890.

material including soil, sand, and rock, the overlook offers a view of a gradually diminishing pit. The Eveleth Taconite Company began dumping overburden and other waste rock in the pit in 1993.

The beauty of all Mesabi Range pits is ephemeral, the changing landscape reflecting past and current mining operations. Advances in science and technology mean that the mine dump of today may well be a mine site of tomorrow—if it becomes profitable to mine "lean" ore with less iron content.

Leonidas Overlook stands above the old Hull-Nelson Mine, originally called the Adams. Portions of the Gross-Nelson and Spruce Mines are also visible. From 1901 to 1978 the Hull-Nelson Mine produced 19 million tons of ore. When Eveleth Taconite formed in the 1960s, the mines here became part of the larger Thunderbird Mine.

You can see remnants of old mining operations from the overlook, including an old mine building and areas in various stages of regrowth, with grass and trees slowly covering sections of the red earth. You also can spot a modern road running through the old mines, the city of Virginia, and the Leonidas location site.

Much of the overburden used to construct the overlook came from the Leonidas Mine, named for Leonidas Merritt, one of seven brothers who discovered the mining potential of the Mesabi Range. The Leonidas Mine began production in 1908 and was one of several early Mesabi mines that initially operated as an underground mine. It closed in 1980.

In its early years, the Leonidas Mine had its own "location," a group of houses and shops built exclusively for the miners and their families. Eventually, this settlement developed into the community of Leonidas under the laws of Minnesota. It is also visible from the overlook.

The development of the Overlook was a joint project of the city of Eveleth and Minnesota's Mineland Reclamation Division Iron Range Resources and Rehabilitation Board.

Directions: From the Twin Cities, follow I-35 north for about 135 miles to State Highway 33 near Cloquet. Take 33 north to U.S. Highway 53 and continue north to Eveleth about 55 miles. The overlook is 1 mile west of Eveleth on County Highway 101.

Season/hours: Late spring through fall, sunrise to sunset.

Length: The overlook provides a 360-degree view and you can see several miles in all directions.

Precautions: There are drop-offs; carefully supervise young children.

Amenities: The view is especially breathtaking in fall.

Contact: Eveleth Area Chamber, 122 Grant Avenue, Eveleth, MN 55734. Phone: (218) 744-1940.
Web site: www.evelethmn.com.

A timber frame marks the spot of Edmund J. Longyear's first drilling site.

Longyear Drill Site

The timber frame points skyward like a four-sided teepee. It doesn't look like much, but from this humble beginning grew the Longyear Exploration Company, which provides core-drilling services around the globe. The Longyear Company has searched for iron ore in Minnesota, Venezuela, and Peru, copper in Arizona and Rhodesia, and zinc and silver in Argentina and China. Its construction programs even included test borings for San Francisco's Golden Gate Bridge.

Edmund J. Longyear arrived on the Mesabi Range in 1890 as a fresh graduate of the Michigan Mining School at Houghton. He set up "housekeeping" in a one-room shack with his bride and, with two partners, set up a steam-drilling rig at this site.

He sank a diamond-bitted drill 1,293 feet into the earth, the first time anyone had used such a drill on the range. The drill produced core samples that could provide information about the existence and location of ore deposits. Although Longyear did not find evidence of usable ore, he did begin a new era in iron ore exploration.

The E.J. Longyear Company began offering its contract services to mine developers. From 1890 through 1911, the company drilled hundreds of samples along the 90-mile Mesabi Range. In 1911 Longyear expanded the company, buying out his partners (one was his uncle) and establishing the Longyear Exploration Company. Over his career, Longyear directed the drilling of more than 7,000 test pits in Minnesota and far beyond its borders.

The original Longyear drilling site is down a quarter-mile fitness trail that begins behind the information booth in the parking lot. In addition to the timber frame above the test pit, the restored site features a drill that would have held the diamond bits to core the earth, a steam boiler, a water pump to cool the bits, and churn buck to drive down and later remove the drill casing.

Walking the wilderness path to the still-remote site duplicates to some extent the experience of E.J. Longyear and his partners. They explored the vast white pine forests of the range, searching—and eventually finding—iron deposits for many mining companies. The success of Longyear's group depended on holding to their dream with the strength of the mineral they sought.

The site is east of Virginia. Take U.S. Highway 53 a few miles south to State Highway 37. Turn east on 37 and drive for about 15 miles through Gilbert to Aurora. In Aurora, turn east on County Highway 100 and go 5 miles to Hoyt Lakes. There, turn left on County Highway 110 and go north for about 3 miles to County Highway 666. Turn right and look for a parking lot on the right with an information booth on the far side.

Contact the Iron Range Historical Society, Box 786, Gilbert, MN 55741. Phone: (218) 749-3150. Web site: www.irontrail.org/longyear.htm.

32
Mineview in the Sky
St. Louis County

THE "KING OF THE LODE" stands guard at the entrance of Virginia's Mineview in the Sky. Dwarfing all vehicles that climb the drive to the parking area, the 240-ton, bright yellow haul truck provides an investigative goal and photo opportunity for most visitors. Still, a photo does little justice to the sheer bulk of the

Tires on the King of the Lode measure about 8 feet in diameter.

machine, looming over inquisitive gear-heads young and old, who envision themselves high in the cab, deep in a pit.

Built for the use of pit foremen, the overlook looms several hundred feet above the canyon that today composes the Rouchleau Mine Group. Use the viewing devices to pinpoint sights about the gorge. Knowledgeable hosts can answer questions about mining equipment, processes, and history in the area.

Miners swarmed to the area following the discovery of iron ore here in 1892 on land leased from the state by Leonidas Merritt and his brothers. By September, developers had platted the town and sold all the lots available. By the end of 1893, John D. Rockefeller had acquired the Merritt holdings, and 15 new mines surrounded the young town.

The Rouchleau Mine, explored by Louis Rouchleau of Duluth, and many of the other nearby mines—including the Lone Jack, Union, Moose, and Enterprise—began as underground shafts. Eventually their maze of tunnels met and intertwined underground. By the 1930s, however, mining here shifted to pit mining, partly because the ground was settling and somewhat unstable and partly because technological improvements made pit mining more feasible. The shafts and tunnels were engulfed in one huge pit.

U.S. Steel and its predecessor companies have owned this mine since 1893. It last produced iron ore in 1977.

For a different perspective of the Rouchleau Mine, head into Virginia to Finntown at the end of Third Street North near Kaleva Hall. A small park on a city bike route offers picnic tables and an observation deck protruding over the pit. The modernistic deck entrance of blue pillars topped by a pink truss provides an intriguing frame to the panorama.

Directions: From the Twin Cities, follow I-35 north for about 135 miles to State Highway 33 near Cloquet. Take 33 north to U.S. Highway 53 and continue north to the town of Virginia, a total distance of about 60 miles. The overlook is on 53 on the south edge of the city.

Season/hours: Open May through September, 9 a.m. to 6 p.m.

Length: The pit is almost 3 miles long and nearly a half-mile wide.

Precautions: Large trucks are mining equipment are on display; carefully supervise young children.

Amenities: Visitor center, playground.

Contact: Virginia Area Chamber, P.O. Box 1072, Virginia, MN 55792. Phone: (218) 741-2717. Web site: www.irontrail.org/rouchleau.htm.

33
Minntac Taconite Plant
St. Louis County

A TOUR OF THE U.S. Steel–owned Minntac plant east of Mountain Iron provides an overview of the sound and scope of modern-day taconite production. While Minnesota's early mines produced millions of tons of hematite, a nonmagnetic iron ore, today's plants focus on taconite, a chert that contains both hematite and magnetite. Named for the Taconic Mountains—a range of the Appalachians covering parts of New York, Massachusetts, and Vermont—taconite is a hard rock containing lower levels of iron than the high-grade ore generally mined in the late nineteenth and early twentieth centuries. In fact, the Minntac process is designed to work with taconite having an iron content below 28 percent.

One word sums up the Minntac operations: big. The plant covers 37,000 acres, including wildlife habitat, as well as two pits. The East Pit is nearly 2 miles wide, 3 miles long, and 370 feet deep. The West Pit is smaller but with a depth of about 400 feet. The tailings basin, containing all the material not used in the pellets, covers 8,000 acres.

Your tour includes a stop at an observation stand above the East Pit, where you watch 240-ton trucks slowly climb the 11 percent grade out of the pit. One reason mining companies switched from trains to trucks is that the trains could climb a grade no more than 1.5 percent, requiring a slow, low-slope spiral of many miles to exit the pit. Besides being more efficient, haul trucks allow more flexibility, enabling plant operators to blend loads from different areas of the pit in order to maintain a constant grade feed to the plant.

Today's diesel-fueled trucks replaced earlier 170-ton diesel-electric combinations that featured an electric motor on each giant wheel. Huge shovels with 36-cubic-yard buckets fill the 240-ton trucks in about five scoops. Working 365 days per year in all weather, the trucks carry 170 loads per day into the plant. Driving one of these truly monster trucks is like "driving a little house," according to one tour guide. The size of the vehicle limits the driver's vision, and he is unable to see a man on the ground until that man is 100 feet away. Each year, the mammoth trucks flatten one or two unmanned, ordinary pickup trucks. The cost of a new truck now exceeds $6 million. The cost of a single tire is $20,000 and each lasts 10 months.

Water trucks keep the dust down on the mine pit roads during dry weather, to improve workers' visibility. At the same time, billions of

The Minntac Mine tour includes a close-up view of an enormous mine truck.

gallons of water are pumped out of the pits annually. This is likely one of many reasons Minntac's monthly electricity bill is $5 million.

Depending on the timing of your tour, you may see blasting, which occurs two to three times per week. First, workers uncover the taconite using shovels and trucks. Then they drill a blast pattern of perhaps 100 holes (20 holes minimum) in the taconite. Each hole is filled with the equivalent of the explosives used in the Oklahoma City bombing. Precise timing and detonating equipment breaks the taconite into chunks that can be hauled to the plant.

Leaving the observation area, your bus enters the complex of buildings that comprises the processing plant itself. The bus circles among the buildings, which include the mine control building. Here workers use computers to track the type and amount of ore in every truckload leaving the pits. All trucks and shovels are equipped with global positioning equipment. Your bus also passes welding and maintenance shops and a series of crushers. The primary crusher extends more than 100 feet into the ground; and the second- and third-stage crushers reduce the incoming ore to pieces less than an inch in any direction.

The bus stops near the concentrator, a half-mile-long building containing rod and ball mills. The crushed ore is mixed with water and ground first in the rod mill, then in two ball mills. When the iron leaves the concentrator it has been ground to the consistency of face powder. With the big drums rolling right and left, conversation is impossible in the noisy concentrator. Even the cement floor vibrates beneath your feet.

Iron particles are separated magnetically from silica (sand) and other components. From the concentrator this dust is pumped into a slurry, from which more silica is removed.

Next, the magnetite concentrate moves to the agglomerator where it is mixed with binding agents, formed into marble-sized pellets, and baked at about 2,400 degrees Fahrenheit. After cooling, the iron-rich pellets—now containing 65 percent iron—are loaded on train cars bound for the steel mills via Duluth and Two Harbors.

About an hour after you enter the plant, your tour bus completes its circuit of the mineral processing operations. Returning to the plant gate, you pass Minntac's fire department, medical facility, and the black, windowless administrative building. Called the "black box" by employees, the sleek, shiny headquarters stands like a monument to the past and future of Minnesota mining.

Directions: From the Twin Cities, follow I-35 north for about 135 miles to State Highway 33 near Cloquet. Take 33 north to U.S. Highway 53 and continue north to the town of Virginia, a total distance of about 60 miles. In Virginia, turn west on U.S. Highway 169 and drive west 3 miles to Mountain Iron. Tours leave from the Visitor Center in downtown Mountain Iron.

Season/hours: Call for dates and times.

Length: This bus and walking tour takes approximately an hour and a half.

Precautions: Wear casual clothes and good walking shoes.

Amenities: Visitors are issued hard hats and safety glasses.

Contact: For tour information, call: (218) 749-7469.
Web site: www.irontrail.org/.

34

Wacootah and Minntac Overlooks

St. Louis County

EAST OF MOUNTAIN IRON, a gravel road leads to the twin Wacootah and Minntac Overlooks. The first offers a view of the water-filled Wacootah Mine pit; the second, higher vantage point looks down on the operations of Minntac's pit. In the distance, the outlines of the plant poke above the horizon. Both overlooks tend to be windy—providing a pleasant experience on a hot summer day.

The Minntac Overlook offers bins of taconite pellets and crude ore samples as souvenirs for visitors. Tread carefully as you approach the bins. Pellets may be scattered on the ground, and walking on them is like walking on marbles!

Directions: From the Twin Cities, follow I-35 north for about 135 miles to State Highway 33 near Cloquet. Take 33 north to U.S. Highway 53 and continue north to the town of Virginia, a total distance of about 60 miles. In Virginia, turn west on U.S. Highway 169 and drive west 3 miles to Mountain Iron. Follow the signs in town along Main Street. Both overlooks are up a gravel drive opposite the main entrance to the Minntac plant.

Season/hours: Open daily spring through fall, from 9 a.m. to sunset.

Length: This overlook is about 1,650 feet above sea level.

Precautions: Walk carefully on the taconite marbles surrounding the bins in the Minntac Overlook.

Amenities: There are no restrooms or other amenities.

Contact: Iron Trail Convention and Visitors Bureau, 403 N. 1st Street, Virginia, MN 55972. Phone: (800) 777-8497. Web site: www.irontrail.org/mtiron.htm.

Viewed from a distance at the Minntac Overlook, the sprawling plant on the horizon is dwarfed by the size of its pit. Evergreen plantings control erosion in the foreground.

Headframes

The headframe standing starkly against the sky at Soudan Underground Mine State Park serves as an historical marker to underground mining everywhere on the Iron Range. Yet several other headframes rise above the horizon of the range. They commemorate a shaft dug, a mine worked, and an industry that helped build the state.

Likely the easiest to find, the Bruce headframe guards the eastern approach to Chisholm, rearing up suddenly on the north side of U.S. Highway 169. For a closer look, walk up the recreational trail bordering the highway. The last headframe on the Mesabi Range, the Bruce headframe once lifted ore up a 300-foot shaft, which is now filled in.

Just north of Ely, the Pioneer Mine headframe provides a focal point on the horizon, with Miner's Lake (a former pit) in the background. Pioneer Mine opened in 1888 and closed in 1967; its history is the history of Ely. From downtown, follow County Highway 21 (Central Avenue) to the north side of Miner's Lake. You also can reach the headframe via the Trezona Trail, a five-mile recreational trail that circumnavigates the lake.

35

Soudan Underground Mine State Park

St. Louis County

WHEN YOU FIRST CATCH sight of the 90-foot headframe looming over the mineshaft and hear the roar of the hoist motor, you may have second thoughts about climbing into the "cage" and descending a half mile into the Soudan Underground Mine. (What? Get into that contraption? I don't think so!) But don't let a moment's hesitation rob you of a unique down-under adventure in Minnesota's oldest and deepest underground mine.

Buy your tickets. Watch the pretour movie. Don your hardhat. And climb bravely into the cage when your turn comes. As it hurtles jerkily downward, you watch the openings of tunnels slip by the window. It feels as if you're dropping at the speed of light, but actually it's little more than ten miles per hour. And although the trip feels like a noisy elevator ride, it's not a straight descent. The shaft is at a 78-degree slant.

Unlike an elevator, the cage has wheels on one side that ride on tracks laid in the side of the shaft, which is hewn out of rock and lined with concrete and iron. Beneath the cage hangs a skip that once carried up to six tons of ore. At the top, the cage is attached to hoist ropes that rise up the headframe and over the giant sheave wheel. From there, the ropes run into the nearby engine house, which contains the huge hoist engine that powers the cage up and down. The engine house is often located at a little distance from the shaft, to ensure that vibrations from the powerful engine do not dislodge rocks in the shaft. The distance is also a fire prevention measure: the engine house may burn, but the fire will not endanger the headframe.

As you descend, look for bats near the top of the shaft and passage openings in the wall. Three or four minutes after you step into the cage, it finally stops at the mine's 27th level, 2,353 feet beneath the surface. There are 50 miles of tunnels in the Soudan Mine and you begin your journey through them via a train, which runs three-quarters of a mile into the tunnel network. Your ride is bumpy and breezy, as wind sails unrestricted through the vented nine-by-nine drift or horizontal passage, constructed in 1951. Along the way, your guide explains that 360 gallons of water pour into the mine every minute and must be continually pumped out.

The historic headframe at the Soudan Underground Mine State Park takes visitors almost a half mile beneath the surface.

The train stops when it reaches the Montana Ore Stope. A **stope** is an underground void created by the extraction of ore. Some stopes are as large as major exhibition halls. To create the stope, Soudan's workers mined upward from a drift, extracting both ore and waste rock. The ore was carried out and the waste rock left in the bottom of the stope, raising the floor as the miners simultaneously raised the level of the ceiling above them. This process is called the cut-and-fill method of mining. In some mines, large stopes must be supported with timbers as a precaution against cave-ins; however, the Soudan is in very hard metamorphic rock, where timbering is not needed.

At this point in your tour, the guide turns on a drill, cautioning you to "cover your ears." The roar is deafening and your guide observes that the miners who once worked here never wore protective earmuffs or dust masks.

The early miners used a tugger (winch) that powered a scraper

Soudan Underground Laboratory

Since the mid–1980s, the Soudan Underground Mine has provided a home to a particle physics laboratory operated by the University of Minnesota–Twin Cities. The region's dense, 3.6-billion-year-old greenstone shields the lab from cosmic rays and other interference found on the planet's surface.

The lab's first experiment is an attempt to detect proton decay. It slows proton decay products as they pass through large corrugated steel sheets and involves a detector using a combination of argon gas, thin copper strips, and gold-plated wires to follow the behavior of decay products. While no lab in the world has so far succeeded in detecting proton decay, the Soudan project has yielded much useful information.

A project now under construction is the Main Injector Neutrino Oscillation Search (MINOS). Its goal is to determine whether a fundamental particle, the neutrino, oscillates, or changes type—moving from muon neutrino to tau neutrino and back again. MINOS is a partnership of the University of Minnesota, 30 other institutions, and Fermilab, operated by the U.S. Department of Energy near Chicago. Fermilab's particle accelerator will send a beam of neutrinos 450 miles through the earth to Soudan, where two **supermodules** will attempt to detect whether the neutrinos have changed type. Each supermodule contains 243 steel plates that are 24 feet wide and weigh 5,400 tons.

Large caverns for these and other planned experiments have been excavated from the mine and are accessed from existing mine passages. The same elevator that carries visitors in and out of the historic iron mine transports lab and construction personnel.

Other underground physics labs are in mines in South Dakota, Ontario, and Japan, as well as in mountains in other countries.

bucket to scrape the loose chunks of ore into a deep hole or chute. The ore was loaded into a **Granby car**—a side-dumping mine car that the miners pushed to the shaft, where the ore was loaded into the skip.

In addition to the ore chutes, "manways" provided access from one level to the next. The miners would climb up and down these vertical passages instead of walking all the way back to the shaft. Today, a spiral staircase provides easier access up one manway.

No one has mined here since 1962. About a year after it closed, U.S. Steel Corporation gave the mine to the state for the development of a park. Although considerable iron remains underground, it's no longer feasible to extract it. During 80 years of mining operations, 15.5 million tons of ore were removed from the Soudan Mine.

Iron was discovered here during the short-lived 1865 gold rush, however, due to a lack of rail transportation and ore refining plants, mining did not immediately begin. In fact, an open pit mine was not established until 1892; two years later Minnesota's first shipment of iron ore left the Soudan Mine and inaugurated the Iron Range. Two pieces of ore from this shipment were saved as memorials: one is on display in the park and the other is at the public library in the Lake Superior port city of Two Harbors.

At first, the Soudan was an open pit mine encompassing as many as seven pits. A short, wooded trail leads from the park road to the first pit dug, and at the end of the trail you can see the cavelike opening of the Breitung Pit. All of the Soudan pits tended to be deep, since the ore body headed almost straight downward into 2.7-billion-year-old metamorphic bedrock. The deep pits were dangerous. Rocks would frequently and unexpectedly fall from the walls, and 125 men died before the mine was converted to an underground operation about 1890.

The early underground miners worked by dim candlelight, often carrying a shift's supply of candles in a Prince Albert tobacco tin, curved to fit comfortably into a back pocket. It was not until 1918 that carbide (acetylene) lamps were introduced into the mines. In 1936 these began to be replaced by electric lamps powered by rechargeable batteries; essentially, these are no different from the headlamp and battery pack assemblies worn by miners today.

In the early days, all work was done by manpower and mule power. The mules were lowered into the mine, where they lived and worked for six months at a time. After their underground tour of duty, the mules got a two-month vacation in the sun.

From the beginning of mining at Soudan, much work took place aboveground, and several buildings remain. The park's ticket office is in the old dry house (or "dry") where the miners would change clothes and hang their wet overalls to dry at the end of a shift. You also can tour the hoist house and see the crusher house and the chutes that once carried crushed ore into waiting rail cars. The rail line can still be seen at the base of the bluff beneath the mining operations and you can walk out onto a platform above these tracks.

If you have the time, hike along the park's trails to catch sight of a variety of hawks, songbirds, and owls. Despite the echoing roar from the hoist engine every 30 minutes on summer days, deer, black bear, timber wolves, and fishers also live in the park.

Directions: From the Twin Cities, follow I-35 north for about 135 miles to State Highway 33 near Cloquet. Take 33 north to U.S. Highway 53 and continue through the town of Virginia to State Highway 169, a total distance of about 65 miles. Turn northeast on 169 and drive for about 22 miles to the community of Soudan. Follow the signs to the park, about three blocks off of 169.

Season/hours: Tours available 10 a.m. to 4 p.m. from Memorial Day weekend through September.

Length: The tour lasts a little over one hour.

Precautions: Wear good walking shoes and bring a jacket or sweatshirt since the mine is about 50 F degrees year-round. The tour includes climbing a short spiral staircase but otherwise is not difficult.

Amenities: Visitor center, movie theater, restrooms, public telephone, picnicking, hiking, fishing, and snowmobiling.

Contact: Soudan Underground Mine State Park, P.O. Box 335, Soudan, MN 55782. Phone: (218) 753-2245. Web site: www.dnr.state.mn.us/parks/soudan_underground_mine/index.html.

36
Ely's Peak Tunnel
St. Louis County

ENTERING ELY'S PEAK tunnel, many visitors grow silent, peering into the gloom ahead and treading with care on the rocky railroad bed. A stiff breeze wraps round them and carries away any words they speak. Just as the light from the tunnel entrance behind them fails, they notice a glow ahead. Although the 469-foot tunnel curves, flashlights aren't necessary to explore it. Standing in the middle of the tunnel, visitors can clearly see both entrances.

Dug out of basalt, a dark volcanic rock, the tunnel was constructed about 350 feet above Lake Superior in the late nineteenth century. It was used by the Duluth Winnipeg and Pacific Railroad until the company abandoned the line in the 1970s. Now owned by the city of Duluth, the old railroad bed is a recreational trail, linking to the Willard Munger State Trail, which begins (or ends) in Duluth. Named for the Minnesota state representative who devoted his career to protecting the environment, the trail is one of the longest paved trails in the United States. It and several other state trails travel through a number of short, concrete tunnels underneath highways, but none compares in magnitude or mountainous beauty to the Ely's Peak tunnel.

If you travel to the tunnel by the Munger or the Duluth Winnipeg and Pacific Trail, you can hike or cycle. Your dog is welcome, too, although it must be on a leash. In winter, you can use snowshoes, cross-country skis, or a snowmobile. The tunnel is some 12 miles south of the Willard Munger Inn, where you can find a trail map, lodging, refreshments, and bicycles to rent. You also can access the old railroad trail from a parking area on Beck's Road. This route involves a 15-minute challenging hike on a steep, rocky trail.

The Duluth Winnipeg and Pacific trail runs through Tunnel Bluff, one of two cliffs (the other is Northwestern Bluff) on Ely's Peak, near the Spirit Mountain area southwest of Duluth. The peak is named for Edmund Ely, an early Presbyterian missionary and pioneer who established the first school in St. Louis County in 1834. Today, rock climbers ascend several routes up these bluffs; most routes involve climbing via numerous vertical cracks in the rock faces. You do not, however, need to be a rock climber to reach the top of the peak, which offers fantastic views of the valley below.

Directions: The trailhead is at Willard Munger Inn in Duluth, across Grand Avenue from the Lake Superior Regional Zoo. You also can access the tunnel from the Beck's Road Willard Munger State Trail parking area. Take Grand Avenue south to Beck's Road (County Highway 3) and turn right. Then, turn right again on 123rd Avenue West, which ends at the parking area. (If you miss this turn, you will intersect I-35 in 2.5 miles.)

From the parking area, go to the Munger Trail. Head up the trail to your right, past a house and driveway. After a short distance, take the trail to the left, which leads across a field. Cross the railroad tracks—but watch for trains, since these tracks are in use today. The area around this crossing is often muddy. Follow the red dirt and gravel trail into a predominantly birch woods. Here, the trail climbs upward, becoming rocky and difficult, but soon intersects with the old railroad bed. You can see the tunnel to your right. If, instead of heading into the tunnel, you cross the old railroad bed, the trail continues to the top of the peak, offering a magnificent panorama.

Season/hours: Year-round.

Length: The curved tunnel is 469 feet long.

Precautions: Getting to the tunnel involves either a 12-mile easy hike, bike, snowmobile, or cross-country ski trip, or it requires a challenging seven- to ten-minute hike up a steep, rocky trail.

Amenities: The Willard Munger Inn offers single and double rooms and studios, some with whirlpools and fireplaces. In winter, the inn has special ski packages; the rest of the year it offers free bicycle use with a room and adventure bike tours including shuttle service.

Contact: Willard Munger Inn, 7408 Grand Avenue, Duluth, MN 55807. Phone: (800) 982-2453 or (218) 624-4814. Web site: www.mungerinn.com.

37 Highway 61 Tunnels
Lake County

ITS ROCKY CLIFFS, pebble beaches, and bold headlands have drawn visitors to Minnesota's North Shore for years. Public and private development, including thousands of acres of parkland, miles of trails, and comfortable lodging, have added to the list of attractions.

This rugged, spectacular shoreline was shaped by millions of years of geologic action. More than one billion years ago, lava flows rose from the area that is now beneath Lake Superior. You can see evidence of these flows in the Sawtooth Mountains that approach the lake in northern Lake and Cook counties and in the basalt rock along the shoreline—for example, at Palisade Head, a mile south of Tettegouche State Park, and at Shovel Point, in the park just northeast of the mouth of the Baptism River. Palisade Head offers particularly awe-inspiring views, and on a clear day you can see the westernmost of Wisconsin's Apostle Islands, about 25 miles away.

Millions of years after the lava hardened, glaciers moved over the area, scouring and carving the outcrops, cliffs, and gorges. Today, streams and rivers in the region move swiftly toward Lake Superior, often whirling through rapids and plunging over waterfalls.

It's no wonder that visitors come to view the power and beauty of this craggy landscape. A weekend spent hiking, fishing, or just plain exploring here can vanquish the most serious case of modern-age stress. Deadlines? Forget them—for a while, at least. Visitors quickly replace any sense of being overwhelmed at the workplace with a sense of overwhelming wonder at the majesty of the North Shore.

All visitors begin their North Shore adventures from State Highway 61, which runs north from Duluth all the way to Thunder Bay, Ontario. Along the way, it traverses towns built for commerce. One hundred years ago, towns like Two Harbors, Tofte, and Grand Marais primarily served the mining, lumber, and fishing industries. Today, of course, visitors also do their part to keep Minnesota's economy healthy.

The approximately 150-mile stretch of 61 from Two Harbors north to Grand Portage is so stunning that it is one of fewer than 20 routes in the United States designated an All-American Road by the Federal Highway Administration. The designation marks highways so scenic

Silver Creek Tunnel extends 1,400 feet through the solid cliff along Lake Superior's North Shore.

that they are destinations in themselves. The route is also one of Minnesota's Scenic Byways.

Driving on this route north from Two Harbors, visitors soon come to the Silver Creek Tunnel, one of two engineering feats designed to make the highway safer for the growing number of travelers who tour here annually. Before the tunnels, Highway 61 contained several sharp and steep curves that sometimes made for hazardous driving. To make matters worse, rockslides along the eroding shoreline occasionally closed the road.

Completed in 1995, Silver Creek Tunnel extends 1,400 feet through the solid cliff. About five miles north, the 854-foot Lafayette Bluff Tunnel also improves safety for North Shore visitors. Both tunnels are convex, curving with the bluff and away from the lake, and they enhance rather than detract from the beauty of the rocky terrain. In fact, the Lafayette Bluff Tunnel received the Seven Wonders of Engineering Award from the Minnesota Society of Professional Engineers.

Thanks to the tunnels, State Highway 61 now gives safe access to the majestic North Shore. In addition to lodges, marinas, and beaches, the road leads travelers to several of the state's most popular parks—Tettegouche, Gooseberry Falls, and Split Rock Lighthouse, to name a few.

Directions: State Highway 61 runs northeast along Lake Superior from Duluth to the Canadian border.

Season/hours: Year-round.

Length: The drive is approximately 150 miles.

Precautions: The drive is accessible to all. Many attractions along its route feature accessible facilities and services.

Amenities: Where to begin? The North Shore is home to magnificent state parks, homey lodges, and recreational opportunities for visitors of all tastes—including antique collectors, fishermen, mountain bikers, cross-country ski enthusiasts, and rock hunters.

Contact: To learn more on the North Shore or Minnesota's Scenic Byways, call the Minnesota Office of Tourism at (800) 657-3700 or (651) 296-5029; or visit its Web site: www.exploreminnesota.com.

38

Gooseberry Falls State Park

Lake County

WALKING ALONG THE paved trail from the Joseph N. Alexander Visitor Center to the Upper Falls in Gooseberry State Park, you don't expect to find a cave. But there is one. At the end of the paved trail, the Upper Falls thunders on your right, and the cave entrance is visible on your left. A short, sloping trail leads up to the cave, which measures about 10 by 15 feet, and is 8 feet high.

This basalt cave contains green lichens and red mud, which may be broken-down rhyolite. The basalt layers in the region frequently contain pockets of rhyolite, another igneous rock.

Seventy years ago, this cave also may have contained groceries. The Civilian Conservation Corps developed the park, and records indicate the CCC used a cave as a root cellar. Park officials believe this is that cold-storage cave.

Gooseberry also contains scores of smaller cavities on the bluff along the Gooseberry River, which you can see by hiking upstream to Fifth Falls. The park contains 18 miles of hiking trails and provides a link to the Superior Hiking Trail, which runs along the North Shore.

Directions: From Two Harbors, follow State Highway 61 approximately 13 miles to the park.

Seasons/hours: Open year-round; however, some facilities are closed in winter.

Length: The walk from the visitor center to the cave is about one mile.

Precautions: The trail to Upper Falls is paved and wheelchair accessible; however, the short sloping side trail to the cave makes it inaccessible to wheelchairs.

Amenities: Camping, picnicking, hiking, trout fishing, and skiing.

Contact: Gooseberry Falls State Park, 3206 State Highway 61 East, Two Harbors, MN 55616. Phone: (218) 834-3855. Web site: www.dnr.state.mn.us/state_parks/gooseberry_falls/index.html.

39
North Shore Sea Caves
Cook and Lake Counties

ASIDE FROM SWIMMING in Lake Superior's frigid waters (not recommended), there are generally two ways you can see Lake Superior's sea caves: visit them by cruise boat or kayak or spot them from certain observation points on the shore. Unlike Wisconsin's Squaw Bay sea caves, which you can hike or ski into during severe winters that freeze the bay, most North Shore sea caves are inaccessible to hikers in any season. The ice here never freezes solid enough to permit a two-legged approach.

There is one exception: in drier seasons and when Lake Superior is calm, you sometimes can enter a five-foot-high cave overlooking Burlington Bay in Two Harbors. In Lakeview Park, the approach is via

In Two Harbors, Lakeview Park offers a dry-land approach to a Lake Superior sea cave. Photo courtesy of Greg Brick.

a gravel road that continues the line of Second Street south into the park. It is also just off the Sonju Harbor Walking Trail. Driftwood and stones cover the floor of the 20-by-20-foot cave.

Adventurers also can safely visit many sea caves by kayak. If you don't own a kayak, several outfitters can help. These include Raven Moon Adventures of Duluth and Cascade Kayaks of Grand Marais, which offer rentals and guided tours of the shoreline's rock arches, palisades, and sea caves. The Lake Superior Water Trail, partially complete, will eventually circumnavigate the lake. Already, campsites and rest areas are available to North Shore kayakers.

Less daring explorers can visit by cruise boat (try Grandpa Woo Excursions in Beaver Bay) or view the caves from several vantage points on land. Look for observation points near bays and inlets or where the shoreline zigzags like a broken pane of glass. These observation stations provide the angle you need to peer into the caves.

One such opportunity is near the mouth of the Baptism River in Tettegouche State Park. Near the entrance, pull into the Baptism River Rest Area parking lot. Follow the trail from the parking lot to the Shovel Point trail that skirts the shore. Three or four vantage points along this short trail provide good views of sea caves and a rock arch. You also can scramble down to a rocky cove, where the sheltering cliffs and expanse of lake and sky just might make you believe that you're the last survivor on planet Earth. Bring your binoculars for the best view of the caves.

Directions: Tettegouche State Park is on State Highway 61, 4.5 miles northeast of Silver Bay.

Season/hours: Year-round.

Length: The Shovel Point Trail at Tettegouche State Park is two miles in length. The Park features more than 20 miles of hiking trails and in winter maintains 12 miles of cross-country ski trails and another 12 miles of snowmobile trails.

Precautions: The Shovel Point Trail is moderately difficult.

Amenities: Tettegouche State Park offers campgrounds and cabins for year-round use. In the summer you can rent canoes to explore the inland waterways, and in winter snowshoes are available.

Contacts: (1) Tettegouche State Park, 5702 Highway 61, Silver Bay, MN 55614. Phone: (218) 226-6365. Web site: www.dnr.state.mn.us/parks_and_recreation/state_parks/tettegouche/. (2) Raven Moon Adventures' Web site: www.ravenmoonadventures.com. (3) Cascade Kayaks, 20 E. First Street, Grand Marais, MN 55604. Phone: (800) 720-2809, (218) 387-2360. (4) Grandpa Woo Excursions, P.O. Box 448, Beaver Bay, MN 55601. Phone: (334) 421-4211.

40
Voyageurs National Park
Koochiching and St. Louis Counties

TO GET THE FEEL of the late-eighteenth- and early-nineteenth-century voyageur experience while you explore an old mining site, sign up for the Gold Mine Boat Tour at the Rainy Lake Visitor Center. As the small cruise boat threads its way among the islands, close to the 2-7-billion-year-old rocky shorelines, you may feel that you're not dressed correctly for the occasion: a voyageur's moccasins, a bright sash at your waist, and a large hat to keep out wind and rain would be more suitable than T-shirt and sandals.

The Little American Mine site contains shafts and this pit on Little American Island in Voyageurs National Park.

Getting to one of the gold mines in the park requires a boat trip from spring through fall; take the cruise service, your own boat, or a rented canoe, motorboat, or houseboat. During winter, ski, snowshoe, or snowmobile to the mines. In any case, you will notice intriguing wildlife and wooded wilderness terrain on your trip.

The park's name commemorates the French-Canadian traders who carried beaver, other pelts, and Native American trade goods between Montreal

Paulson Mine

Camps, abandoned large equipment, and other remnants of the short-lived Paulson Mine still lie partially hidden in the woods along the Gunflint Trail. The drivable trail (County Highway 12) runs northwest from Grand Marais, and the former mine site is roughly 45 miles from the village.

The Paulson, an underground iron mine, opened in 1888 but closed following the depression of 1893. After it closed, the area was the site of considerable logging activity. The lumber companies used the Port Arthur, Duluth & Western Railway line, constructed from Thunder Bay, Ontario, to the Paulson Mine on Gunflint Lake.

and the Canadian Northwest by canoe. Their route included the waterways between Lake Superior and Lake of the Woods. Voyageurs National Park, established in 1975, contains more than 30 lakes, as well as bogs, marshes, and beaver ponds. Access these waters from entry points at three visitor centers, where you can leave your car and switch to water or ice route transportation.

Since well before the 1849 California Gold Rush, prospectors scoured the United States in search of precious minerals. Gold was found in the Lake of the Woods in the late 1870s, and in 1893 George W. Davis discovered gold in a vein of quartz at Little American Island.

Today there are still veins of quartz containing gold in a belt of **schists** extending from Little American Island about 50 miles west and more than 120 miles east into Canada. (Schists are metamorphic rocks characterized by parallel layers, which sometimes split into flat fragments.) Currently, the gold contained in this belt cannot be profitably mined, and profitability was a challenge in the early days too.

The gold boom launched by Davis's discovery lasted only until 1898. But during those five brief years, men—and a few women and children—flocked to the region. Shafts were sunk, tramways and crushing mills developed, and towns created. Spurred by the Depression of 1893, which left many people out of work, and by the lure of wealth and adventure, they headed north. Many hoped to work in the mines; others planned to offer whiskey, poker, and other amenities to the miners in the new boomtowns.

The Little American Mine was originally owned by Davis and by Charles Moore, who had grubstaked Davis and earlier developed a mine on Lake of the Woods. But mine development was a costly undertaking, so Davis and Moore sold their claim to Jeff Hildreth, who brought in additional investors and dug an **adit**, or horizontal passage into a hillside on the island. He then built a mill at Rainy Lake City to crush the quartz and sank a vertical shaft. When the shaft reached a depth of 43 feet, miners hollowed out the first stope at the

35-foot level. Eventually, miners extended a second shaft to a depth of more than 200 feet.

Yet, the mine turned a profit only briefly, overcome by a series of obstacles and under-capitalization. It changed ownership frequently, as one investor after another sunk funds into the development of the Little American. One year, a worker let $10,000 worth of gold slip into Rainy Lake. Another year, a loading dock collapsed, sending an entire season's work of ore into the lake. Finally in 1898, the sheriff seized the mine, and later attempts to open it ended in failure as the Klondike Gold Rush attracted miners and mining capital to the Alaskan frontier.

The Gold Mine Boat Tour stops at Little American Island, where you can disembark and follow a path to historic mining sites, including the shafts, which have been filled with sand for safety. You can also see an open pit and signage along the way explaining early mining operations.

Back on the boat, the tour continues, passing other islands where you can see buildings that mark the site of an old mining village and the entrance to the adit of Bushyhead Mine on Bushyhead Island. This tunnel extends halfway through the island to a vertical shaft, now sand-filled. The tour boat also motors past Powder Island, where the early miners stored blasting powder at some distance from their mining operations—as a precaution against lightning strikes.

The tour glides by two rocky, treeless islands in Canadian waters that are rookeries for gulls and cormorants. You will hear their cacophony long before you can clearly see these birds. You may see goshawks, herring gulls, and perhaps a loon with a baby on its back. Bring your binoculars and look for "golf ball" bald eagle heads in the treetops.

Other tours leaving from the Rainy Lake Visitor Center take you close to an eagle's nest, historic fishing camp, and to the Kettle Falls Hotel, accessible only by water. Four Rainy Lake resorts also offer lodging and boat rentals.

Directions: From International Falls, follow State Highway 11 east for 10 miles. Turn right on Voyageurs National Park Road and follow it for 1.5 miles to the Rainy Lake Visitor Center.

Season/hours: Year-round. The Gold Mine Boat Tour is offered twice weekly during the summer.

Length: The Gold Mine Boat Tour takes one and a half hours.

Precautions: If you travel by private motorboat, houseboat, or canoe, heed local weather forecasts. Storms can come up quickly. If you travel in winter by skis, snowshoes, or snowmobiles, beware of slush and unsafe ice conditions, as well as blizzards.

Amenities: The path on Little American Island is wheelchair accessible. Voyageurs National Park offers more than 175 campsites accessible only by water. There are also recreational trails; some can be reached only by water. Rainy Lake Visitor Center is open all year. Ash River and Kabetogama Lake Visitor Centers are open during the summer. All offer books, maps, navigational charts, films, and exhibits. Summer naturalist programs provide information on the wildlife and history of the area.

Contact: Voyageurs National Park, 3131 U.S. Highway 53, International Falls, MN 56649-8904. Phone: (218) 283-9821. Web site: www.nps.gov/voya. For boat tour reservations leaving from Rainy Lake Visitor Center, call (218) 286-5470.

41

Isle Royale National Park

Houghton County, Michigan

ALTHOUGH ISLE ROYALE National Park is really a part of Michigan, it lies closer to Minnesota: situated in Lake Superior, it is less than 20 miles southeast of Grand Portage, as the crow flies. You can reach the western edge of the island in two hours by cruise boat. Isle Royale is also accessible by private boat or seaplane. No motor vehicles, bicycles, or pets are permitted on Isle Royale or the smaller islands surrounding it. For overnight stays you can camp or stay at the Rock Harbor Lodge.

Isle Royale contains a small cave, the remains of many nineteenth-century copper mines, and pre-Columbian mining pits. For thousands of years before European explorers arrived on the scene, early Native Americans mined copper using hammerstones to dislodge chunks and hammer them into tools.

The bedrock of the archipelago is basalt, formed more than one billion years ago, when a huge rift appeared in the earth's crust and multiple lava flows left behind a series of ridges. Signs of more recent glaciation remain in deposits in the southwest section of the island.

A 40-mile trail runs along the most prominent ridge, Greenstone Ridge, which traverses almost the length of the island. Side trails lead to abandoned mines, lakes, and lookouts. You can also reach old mining sites by National Park boat service from the visitor centers at Rock Harbor and Windigo. Park rangers lead regular boat/hiking tours to Minong Mine, Suzy's Cave, and other sites during the summer. Since transportation is by foot or boat, allow at least a long weekend to explore Isle Royale.

Directions: The most popular way of getting to Isle Royale from Minnesota is via the boats of the Grand Portage–Isle Royale Transportation Line. They leave from Grand Portage, 153 miles northeast of Duluth on State Highway 61. (See Contacts below.)

Season/hours: Open from mid-April through October 31. Full transportation services available from mid-June through Labor Day.

Length: The park archipelago is 45 miles long. The hike to and from Suzy's Cave from Rock Harbor is 3.8 miles.

Precautions: Many hiking trails are quite challenging. The temperature is cool: the average daily high in July is 68 degrees; the average daily low in July is 50 degrees. As you hike and camp about the island, water must be boiled or treated. Be prepared for mosquitoes, biting flies, and yellow jackets. Bring medications and a first aid kit. No cars, bikes, or pets allowed.

Amenities: Hiking, kayaking, boating, scuba diving.

Contact: Grand Portage–Isle Royale Transportation Line, Inc., 1507 N. First Street, Superior, WI 54880-1146. Phone: (715) 392-2100. Web site: www.GRAND-ISLE-ROYALE.com. Or visit: www.nps.gov/isro/.

42

Amethyst Mine Panorama
Ontario, Canada

AT THE END OF A gravel road about a half hour drive northeast of Thunder Bay, Ontario, is one of the world's largest amethyst mines, which annually produces 95 percent of the amethyst mined in North America. One of several amethyst pit mines in Ontario, Amethyst Mine Panorama offers grand views, informative tours, and a chance to dig your own finds to take home—for a very reasonable price.

The mine was discovered during a road-building project and has operated continuously since 1960. The amethyst crystals, a type of quartz, range in color from deep purple to lilac and lie near the surface.

A tour of the mine involves a walk of about 400 feet on paths and wooden ramps, to a depth of 35 feet. Using explosives, miners break off blocks of rock and then collect the most valuable amethyst

Visitors can wash their found treasures at Amethyst Mine Panorama. Photo courtesy of Steve Lukinuk.

chunks. The waste rock, consisting of 20 percent amethyst usually of a lesser quality, is hauled to a five-acre plot which visitors are welcome to mine.

Occasionally you will find a truly outstanding piece, but you will always find good pieces for your collection or to craft into bookends, jewelry, or other items. The picking area contains 40 years of waste rock and is continually refreshed, according to Steve Lukinuk, who operates the mine with his wife, Lorna, and son Tim.

Even though approximately 40 percent of its 35,000 annual visitors come from Minnesota and Wisconsin, Amethyst Mine Panorama is relatively unknown in Minnesota. Yet for rock hounds, it's well worth the seven-hour trip north from the Twin Cities.

Amethyst from the mine is hauled to Thunder Bay, where it is cut, polished, and crafted into a wide array of items—jewelry, light catchers, key chains, clocks, carvings, bookends, trophies, and pen sets, to name a few. The manufacturing area is behind the retail shop at the Amethyst Gift Centre, housed in a former bank. Before or after you shop, take a short tour of the processing areas.

Directions: From Duluth follow State Highway 61 northeast to Canada Highway 11/17 in Thunder Bay, Ontario, a distance of 185 miles. Turn right on 11/17 and drive 25 miles northeast to East Loon Road. Turn left on East Loon Road and drive 6.5 miles to the mine. The Amethyst Gift Centre is at 400 Victoria Avenue in Thunder Bay. Drive east on Business Highway 11/17 Business (Arthur Street) to May Street. Turn left on May and drive to Victoria Avenue. Turn right on Victoria.

Season/hours: The mine is open daily from May 15 through October 15. Hours: 10 a.m. to 7 p.m., July through August; 10 a.m. to 5 p.m., May 15 through June 30 and September 1 through October 31. The Amethyst Gift Centre, including the manufacturing tour, is open year-round, Monday through Saturday, 10 a.m. to 5 p.m.

Length: A tour of the mine lasts 35 minutes, but allow at least an hour or two for the tour plus time to dig for amethysts. A tour of the manufacturing processes in Thunder Bay lasts ten minutes.

Precautions: The mine tour is partially wheelchair accessible.

Amenities: Other Thunder Bay attractions include the Sports Hall of Fame, Kaministiquia Heritage Park, and the Thunder Bay Historical Museum.

Contact: Amethyst Mine, 400 Victoria Avenue East, Thunder Bay, Ontario, Canada, P7C 1A5. Phone: (807) 622-6908. Web site: www.amethyst-mine.com.

Section IV
Black Hills South Dakota

SPEND A WEEK in the Black Hills and you can visit more significant caves, mines, and tunnels than is possible in a month in many states. In an area stretching about 115 miles from north to south and perhaps 70 miles from east to west, the center hills contain as many as 200 caves. Nine large caves, nine tunnels, and three historic underground mines are open to the public.

Two caves are among the world's longest: Jewel Cave has about 128 miles of mapped passages and Wind Cave is close behind with roughly 103 miles of passages surveyed. Each year, teams of cave explorers discover additional passages, following air currents and pushing into leads off the known passages and chambers. Wind, Jewel, and other caves in the region still have considerable potential for new passages, to add to the maze of underground labyrinth hidden beneath the surface.

Why are there so many extensive caves in the Black Hills? The caves exist in a layer of Pahasapa limestone that circles the inner core of the hills like an oblong racetrack. A deposit of Minnelusa sandstone overlays the limestone. Both of these layers were created more than 300 million years ago when a shallow sea covered the region.

Sixty million years ago the mountains were pushed up from underneath, forming the Black Hills. The Minnelusa and Pahasapa layers in the highest, center area of the hills eroded away, exposing granite, schist, and other Precambrian rock. The Pahasapa layer tilts slightly away from this higher central core.

The publicly accessible caves offer a range of caving activity. A few caves are partially accessible to individuals in wheelchairs and young children. Several offer a menu of tours ranging from relatively easy to strenuous, and a few offer wild cave tours, requiring much stooping, crawling, and steep climbing. Wind Cave, for example, offers five tours—from an easy one-hour tour involving only 150 stairs to a four-hour wild cave tour involving much squirming and crawling. Many caves showcase speleothems rare in Minnesota, as well as the more common stalactites, stalagmites, popcorn, and flowstone.

Caves are not the only underground attraction. Historic gold mines dot the Black Hills, dating from the 1870s when hordes of miners flooded the area in search of instant wealth. Gold is still mined here, although many underground operations have shut down. Three of these sites are open for tours.

This mountainous region also spawned the development of tunnels, first to provide railroad access and later to bring tourists to the area's peaks, wilderness, and underground attractions. Today you can walk, or ride a bicycle or a horse through four old railroad tunnels on the George S. Mickelson Trail, and you can drive through five tunnels on the Peter Norbeck Scenic Byway. Norbeck, a former South Dakota governor, designed this route personally on foot and on horseback, to provide visitors with a personal encounter with the majesty of his beloved region.

Even if your object is to exclusively explore the underground mysteries of the Black Hills, you will inevitably encounter its wildlife. Keep a watchful eye for mountain goats, bighorn sheep, deer, pronghorn, burros that beg food along the highway, and buffalo that know they own the road. Oh, yes—and the prairie dogs that pop up and down in the extensive towns they've built in Custer State Park. ■

43

George S. Mickelson Trail

Lawrence, Pennington, Custer, and Fall River Counties

FOLLOWING THE OLD Burlington Northern Railroad line from Deadwood to Edgemont, the George S. Mickelson Trail accommodates hikers, bicyclists, and horseback riders. It features maximum grades of four percent, more than 100 bridges, and four tunnels. Known as Tunnel A, Tunnel B, Tunnel C, and Tunnel D, they are along a 12-mile stretch beginning north of Rochford to about 2 miles south of Mystic.

Blasted through rock, the tunnels offer shade and a welcoming coolness on a hot summer day. They range in length from 166 feet to 270 feet. A flashlight will help guide your path if you visit on foot.

George S. Mickelson, late South Dakota governor, supported the development of this Black Hills trail, which was completed in 1998. An annual Trail Trek commemorates its completion each September.

Directions: The trail's northernmost point is the Deadwood trailhead near the intersection of U.S. Highways 14A, 85, and 385. The trail follows 85 through Lead, then generally follows County Highway 17 south to County Highway 231 at Rochford. From Rochford, it continues to run near 231 until 231 again intersects with 17. The trail generally follows 17 to Hill City. From Hill City it follows U.S. 385 to Custer. Leaving Custer, the trail follows 385 and State Highway 89 to Pringle. From Pringle, the trail follows 89 to its junction with U.S. Highway 18. The trail loops far to the east of 18 before both come together again north of the southernmost trailhead in Edgemont. The trailhead at Mystic provides closest access to the tunnels. Walking south from Mystic, you come to Tunnel B in 1.7 miles.

Season/hours: Dawn to dusk, spring through fall.

Length: The trail runs 114 miles, from Deadwood to Edgemont.

Precautions: Some sections of the trail traverse private land, and rattlesnakes and poison ivy can be found nearby, so it's advisable to stay on the trail. Bicyclists ride on the right side of the trail and yield to pedestrians. Bicyclists and pedestrians yield to horses. Parts of the trail run through open range areas, where you may encounter free-roaming cattle. Try not to spook the cattle to avoid having them run into another group of trail users. Bicyclists should not pass in the tunnels.

Amenities: Thirteen trailheads provide access to the trail and trail passes. Seven of these offer water, telephones, camping, lodging, and dining. Bicycle rentals are available in Deadwood, Spearfish, Custer, and Hill City.

Contact: Black Hills Trail Office, 11361 Nevada Gulch Road, Lead, SD 57754. Phone: (605) 584-3896. Web site: www.mickelsontrail.com.

44

Broken Boot Gold Mine

Lawrence County

THE TICKET OFFICE, gold panning shed, and entrance to the Broken Boot Gold Mine cling to the hill above the traffic on U.S. Highway 14A outside historic Deadwood. Established in 1878, when the lure of gold launched a flood of miners into Deadwood Gulch, the mine operated until 1904. Half-hour tours showcase the late-nineteenth-century mining operations, which involved picks, shovels, black powder, and strenuous hand labor accomplished by candlelight.

Follow the main tunnel about 850 feet into the mountain to a large stope where you can see many of the tools used by the early miners. You learn how the mining was done, the ore processed, and how the miners lived in the rough-and-tumble western town barely on the edge of civilization.

Directions: From Deadwood, follow U.S. Highway 14A west to the mine just outside of town.

Season/hours: Open daily, 8 a.m. to 5:30 p.m., mid-May through mid-September.

Length: Tours take about one half-hour.

Precautions: Stairway to entrance; easy walk once inside the mine.

Amenities: Gold panning, souvenir stock certificates.

Contact: Phone: (605) 578-1876.

45

Bethlehem Cave

Meade County

THE ENTRANCE to Bethlehem Cave is through a manmade tunnel on the side of a canyon wall. Soon after entering the tunnel your helmet light picks up one of the distinctive features of this cave. The tunnel leads to St. Benedict's Grotto and the Shrine of the Nativity, featuring a rock altar and seating hewn from the rock. Conception Abbey, Order of St. Benedict of the Catholic Church, received the cave as a gift in 1952 and owned it until 1968.

Monks created the Shrine of the Nativity and changed the name of the former Crystal Cave to Bethlehem Cave. The natural entrance is from this first room, up a twisting passage in the ceiling. Looking out from this passage at night, you can see the North Star, which reminded the monks of the star of Bethlehem. They built steps in the natural entrance passage, naming it the North Star Stairway.

Except for the developments in these first rooms, the monks made few changes to the cave and worked to preserve its speolothems and environment.

In 1878, miners began digging the Broken Boot Gold Mine outside of Deadwood, South Dakota.

Two brothers, Adolphus C. and Charles Frederick McBride, discovered the cave in 1889. They were searching for mining prospects and according to one account also hunting mountain lions. They noticed that when they threw rocks down the canyon, the rocks soon disappeared inside the mountain. Exploring the area, the brothers discovered the cave they named Crystal and worked to open it for tours.

Proprietors working with the brothers in the early 1890s cut formations from the cave to use in creating a model cave at the 1893 Columbian Exposition in Chicago. They built the cave underneath the Horticultural Building, dividing a 60-by-80-foot space into seven rooms and filling them with stalactites, stalagmites, flowstone, cave pearls, dogtooth spar crystals, and other formations taken from the cave. At the 1900 Pan-American Exposition in Buffalo and the 1904 Louisiana Purchase Exposition in St. Louis, Adolphus McBride displayed crystals taken from the cave.

Despite these atrocities, the cave remains well preserved, thanks in part to the conservation efforts of the monks and to real estate investor Loui Storm, who purchased the cave from Adolphus McBride, developed tour routes, and successfully promoted the attraction. In his 70s, he deeded it to the monks, hoping it would be used for youth gatherings and retreats. The Abbey transferred ownership to The Shrine of the Nativity, Inc., which in turn passed ownership of the cave to the Loui Storm Foundation.

Throughout these changes in ownership, tours and cave develop-

Homestake Visitor Center

Although the Homestake Gold Mine closed in 2001, its nonprofit Visitor Center in Lead remains open from Memorial Day through Labor Day, offering exhibits, a video, a look at the original open cut claim, and tours of the surface mining operations of the Homestake Mine.

Its history dates to the discovery of an outcrop or "lead" of gold ore in 1876 and encompasses the mining of the open cut and later the development of underground operations. The success of the mine spurred the growth of the towns of Lead and Deadwood and led to the creation of banks, hospitals, and mining-related industries.

When the price of gold increased in the early 1980s, Homestake decided to expand its original open cut, requiring the relocation of 97 homes and a city park, as well as the town's Main Street.

Contact Homestake Visitor Center, 160 W. Main Street, Lead, SD 57754-0887. Phone: (888) 701-0164 or (605) 584-3110. Web site: www.homestake-tour.com.

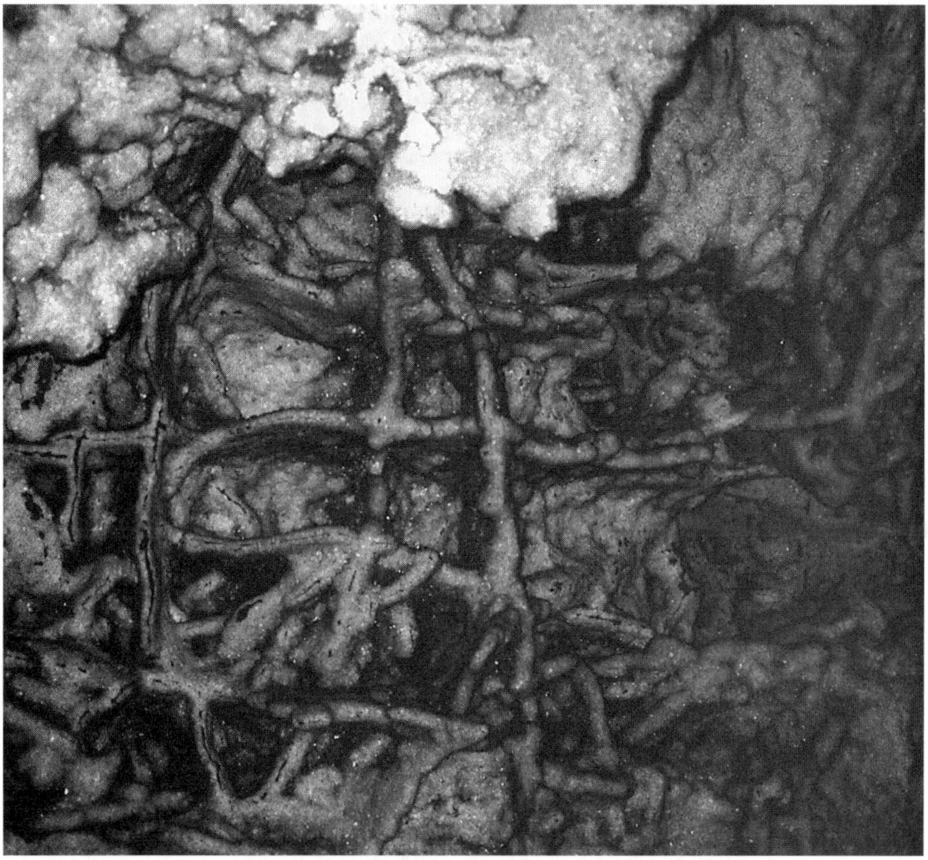

Bethlehem Cave is one of several in the Black Hills that contain boxwork.

ment continued. Though robbed of many formations, there are still plenty of crystals to see, as well as **boxwork**—a calcite formation resembling a honeycomb grown wild, with many compartments of angular, irregular shapes. In the Milky Way, crystals twinkle in the light from your headlamps, and structures like the Shark Pit, Cocoran's Art Gallery, and Wine Cellar are realistic enough to suggest a human creator.

Many visitors continue to view Bethlehem Cave as a God-given wonder. Weddings, Christmas services, and other events are held here annually.

Directions: Take I-90 to exit 40 (9 miles southeast of Sturgis and 17 miles northwest of Rapid City). Follow the signs to Bethlehem Road, and go 5 miles west on Bethlehem Road to the cave.

Season/hours: Year-round. Open September to June, 9 a.m. to 6 p.m., and 9 a.m. to 9 p.m. in July and August.

There are no set times for tours; tours leave when a group arrives and is ready to go.

Length: Bethlehem Cave offers several tour options. One lasts less than one hour; another takes more than two hours. Overnight and spelunking tours are also available.

Precautions: The one-hour tour includes some steps, which are easily manageable for most children and adults.

Amenities: The full name of this attraction is Bethlehem Cave Wild Adventure Park and Lodge, which provides a clue as to the activities possible. Located on 180 acres surrounded by National Forest lands, Bethlehem offers canyon tours, horseback riding, a rock climbing school and climbing tower, snowmobiling, cross country skiing, and hunting. Campsites and cabins, suites, and rooms are available.

Contact: Bethlehem Cave, P.O. Box 592, Black Hawk, SD 57718. Phone: (605) 787-7500. Web site: www.bethlehemcave.com.

46 Beautiful Wonderland Cave

Meade County

IF YOU WANT to experience the difference between a dead and a living cave, tour Wonderland Cave, which has two levels—a higher, drier, dead level and a lower, wetter, living level. The upper section even contains a "touching gallery" where you can feel formations that are no longer growing.

Walk to the bottom of a stairway, and the cave visibly changes, becoming wetter, with formations that are still dripping and ever so slowly growing. The lower level is 280 feet beneath the surface, and rainwater continues to leach into its rooms and passages.

The cave presents a great diversity of formations, for example, stalactites, stalagmites, columns, an abundance of dogtooth spar crystals and cave popcorn, calcite lily pads, and boxwork. Visitors have given the usual creative names to many formations, and your guide points out the White Forest of stalactites and stalagmites, Ice Cream sandwiches, the Queen's Jewel Box, and the Carrot Patch growing in a knee-high, kid-sized chamber. The 40-foot Icicle Fence is one of the

Beautiful Wonderland Cave formations include stalactites, stalagmites, columns, and flowstone (foreground).

more memorable structures.

One of Wonderland's attractions is the variety of features to observe. In addition to the varied formations already named, the cave contains two flowstone "waterfalls," several streams, a man-made wishing well, and seven natural vents. Thanks to the vents, a number of bats make their home in the cave. And thanks to the artificial lighting made necessary by numerous visitors, you can see moss growing in a few sections.

Wonderland was discovered in 1929 on land that is now privately owned but surrounded by national forest preserves. Two loggers, Pete and Thurston Lein, were searching for a lost dog when they stumbled into Wonderland's upper cavern. At first, they tried to mine gold and silver in the cave, but when the effort did not pan out, they opened Wonderland to tours beginning in 1930. During the Great Depression, the Civilian Conservation Corps assisted with the development by building steps and smoothing out passages.

Today, more than two dozen rooms are open to visitors, but several caverns have yet to be explored. If you could repeat a caving trip to the region every few years, there would always be new rooms and passages to experience at Wonderland.

Directions: **From I-90 at Exit 32 in Sturgis, go south on Vanocker Canyon Road to County Highway 137,**

following the signs to the cave; total distance from the exit is 17 miles.

Season/hours: Open daily from May 1 to November 1.

Length: Allow an hour for the mile-long tour.

Precautions: Several short flights of steps. Temperature averages 47 degrees.

Amenities: Picnic area, short trail to the canyon. Across the canyon, you can see the opening of Bethlehem Cave high on the opposite cliff wall.

Contact: Beautiful Wonderland Cave and Natural Park, Box 83, Nemo, SD 57759. Phone: (605) 578-1728 (summer only).

47
Stagebarn Crystal Cave
Meade County

ALTHOUGH NEVER used as a storage facility for stagecoaches, Stagebarn Crystal Cave is set deep in a canyon that once had a stage stop at its mouth. From the stage stop, miners traveled to old workings on a narrow-gauge rail line. The tracks were torn up long ago, but the rail bed still serves as the road to the cave. In fact, a railroad surveyor, Rudolph Stoll, discovered the cave, and it remains in his family today.

You can see the cave entrance from the tiny visitors' building. It's hard to believe that your guide will expect you to climb the steep outdoor stairway clinging to the canyon wall. But the tour leaves as soon as a group of visitors arrive, and soon you find yourself on the staircase to heaven, trying not to look down. More than a hundred steps later—and 90 feet higher—you reach the natural entrance.

Once inside the cool cave, you can relax. There are a few steps, low spots, and damp places, but nothing you encounter in the cave compares to the adventure of reaching the entrance.

Calcite crystals dominate this cave, covering 80 percent of the walls. A few solution cavities, like shallow domes, punctuate the ceiling. One of these, Capital Dome, is said to be the largest in South Dakota. Stagebarn Crystal Cave contains no really large rooms, but some ceilings soar to 30 or 40 feet.

Several deep, narrow canyons trace their origins to joints in the

Twisted columns adorn this room in Stagebarn Crystal Cave.

rock layers, similar to, but smaller than, the canyons in Minnesota's Niagara Cave. These narrow passages lead to a number of wider, shorter rooms filled with stalactites and stalagmites. Along the way, notice a few shell fossils, flowstone, popcorn, and draperies.

The tour descends to the second level of the cave's three levels via a rock spiral stairway—definitely an improvement over a ladder for traversing the steep passage. On the lower level boxwork and shelf work decorate the walls. You also see helictites blooming from stalactites at odd angles and twisted columns, as well as cave ice or mineral rafts. Other notable features include:

- A Tee Pee Town constructed of stalagmites.
- Flowstone Falls.
- Nature's Painted Room, embellished with yellow, gold, and rust.
- Candlewax Falls, where stalagmites appear to have candle wax dripping down their columns.
- The General Store, featuring a little of everything—helictites, stalagmites, stalactites, flowstone, shelf work, and boxwork.

At about six miles, Stagebarn is the third longest cave in the Black Hills, and the sections of the cave not on your tour all contain well-preserved speleothems. Wind Cave and Jewel Cave, of course, are the two longest of the hills' roughly 200 known caves.

Directions: Take I-90 to Exit 48 (17 miles southeast of Sturgis and 11 miles northwest of Rapid City). At the Exit 48, follow Stagebarn Canyon Road and watch for signs to the cave.

Season/hours: Open daily, 8 a.m. to 7 p.m., Memorial Day through Labor Day.

Length: This cave is about six miles long. The one-half-mile tour takes less than an hour.

Precautions: The tour begins with 106 steps (90 vertical feet) up the bluff to the cave opening and includes additional steps in the cave. Temperature averages 44 degrees.

Amenities: Stagebarn Canyon is also known as Botany Canyon, for its diversity of plant life.

Contact: Stagebarn Crystal Cave, 10829 Stagebarn Road, Piedmont, SD 57769. Phone: (605) 787-4505.

48
Crystal Cave Park
Pennington County

CRYSTAL CAVE in Wild Irishman's Gulch provides perhaps the most knowledgeable and authoritative tour of the commercial Black Hills caves. All guides are members of the National Speleological Society, versed in the geology of the hills and the origination of cave formations. Compared to other tours, Crystal circuits emphasize science and downplay the fantastical. If you want to stretch your imagination by searching for animal shapes among the formations, other caves offer better opportunities. But if you want to understand how caves form and speleothems develop, Crystal Cave Park is the place to come.

The smallest of the Black Hills' commercial caves, Crystal contains a good variety of formations. You can see six-pointed dogtooth spar crystals and hailstone spar crystals, as well as the usual stalactites and stalagmites. There is petrified moss and coral and unusual birdbath formations, seen in only one other U.S. cave. (At Stagebarn Crystal Cave—see page 138—you may see a shape termed the Bird Bath, but it is not a true birdbath formation, created over eons by water dripping and minerals precipitating in just the right conditions. The Bird Bath at Stagebarn formed when a chunk of rock fell from the ceiling.)

Discovered by white men in 1880, Crystal Cave was long known to Native Americans, and the battered skull of a Mandan Indian boy resting in one small room proves the point. The Mandan lived in what is today North Dakota but sometimes hunted as far south as the Black Hills.

Originally—and somewhat confusingly—named Wind Cave, the cave was later called Diamond Crystal Cave; postcards with this moniker still exist in the gift shop. The cave was opened for tours in 1931, following the installation of the entrance stairway. That same year, the federal government outlawed creating graffiti in caves. Still, you can find limited graffiti in Crystal Cave. Some walls are also decorated with telltale signs of flooding. Many Black Hills caves were inundated when a dam broke in 1971, necessitating a huge volunteer clean-up effort.

In addition to the water marks, graffiti, and formations, Crystal Cave contains sphagnum moss under the artificial lighting and cave mushrooms. A pack rat has even set up housekeeping in an eye-level niche not far from an opening to the surface.

Walking through Crystal, its passages and rooms feel smaller than those in the larger caves, as if all space in the cave is proportional to its overall size. Its dimensions also contribute to Crystal's easy accessibility and popularity; and all of these factors may affect its temper-

Crystal Cave contains this unusual birdbath formation, beneath a stalactite.

ature, which averages above 50 degrees Fahrenheit. Yet the cave extends at least 300 feet beneath the surface and still contains unexplored areas.

Near the end of your tour, stop at a black light exhibit where certain rock specimens fluoresce purple, yellow, and green, according to their mineral content. Rocks containing fluoride, for example, fluoresce with an alien-green aura—suggesting an intriguing experiment to try at Halloween. After the kids brush their teeth, ask them to stand in front of a mirror and then shine a black light in their direction.

Directions: From Rapid City, drive west 1 mile on State Highway 44 to Cave Road. Follow this 2 miles to the cave.

Season/hours: Open daily, 8 a.m. to 8 p.m., June through August. Open 9 a.m. to 5 p.m., in May, September, and October.

Length: This tour is an easy 45-minute walk.

Precautions: Jacket recommended. This is a relatively easy walk, for a cave. There are 58 steps into—and out of—the cave.

Amenities: The park features a nature trail and a replica Indian village.

Contact: Crystal Cave Park, 7770 Cave Road, Rapid City, SD 57702. Phone: (605) 342-8008.

49

Black Hills Caverns

Pennington County

LEGENDS ABOUND in Black Hills Caverns, which were known to Native Americans for centuries and rediscovered by Calvin Meade, a prospector, in 1882. Walking in Wild Irishman's Gulch, Meade felt a cold wind issuing from an opening at the base of the cliff. This shallow opening is still visible, but today you enter the caverns from an easily accessible man-made opening.

Not far from the entrance is the Moonshine Room, where Meade supposedly operated a still. Legend has it that Meade's moonshine blinded some people and one day angry customers came after him. Meade escaped through a particularly tight, steep passage to the surface. Looking at the small room and twisting chimney, this story seems almost believable.

Packsaddle tours began in the caverns in 1939. Today, you choose between an easy half-hour tour and a one-hour tour that journeys 225 feet beneath the surface to all three levels of the cave. Crystals are abundant on both tours; however, you see the greatest variety of formations on the longer tour—stalagmites, stalactites, soda straw stalactites, helictites, boxwork, cave flowers, dogtooth spar crystal, and nail-head spar crystal, to name a few. Of particular interest is the rare frost crystal.

Along with the crystals and other formations, many shapes are named through the imagination of earlier visitors. The Buffalo Room features, naturally, a crystal buffalo, and you can't miss the fierce alligator guarding another passage.

While a natural stream runs through the lower level, most of the upper cave is dry and dead, and the growth of its formations has ceased. Tours here tend to be drier than at other caves; you're less likely to feel water dripping on your head or down the back of your neck.

In the Lion's Den, the local Lions Club supposedly initiated members by leaving them here with two wooden matches. Eventually, according to the story, the new members got wise and wore layers of old clothing that they lit to use as a torch to find their way out.

Thankfully, you have both light and a guide. Yet the story is a reminder of the caving commandment: always carry three sources of light.

Directions: From Rapid City, drive 4 miles west on State Highway 44 to Cavern Road. Follow this road to the cave.

Season/hours: Open daily, 8 a.m. to 8 p.m., June 15 through August 31. Open 8:30 a.m. to 5:30 p.m., May 1 to June 15 and September 1 through October 15.

Length: The easier, one-half hour Crystal Tour covers 1,500 feet of passages and rooms. The moderately difficult, one-hour Adventure Tour traverses .75 miles.

Precautions: The cave is dimly lighted in spots to aid in preservation of this natural resource. Watch your step.

Amenities: Panning for fossils, minerals, and gems to identify and take home.

Contact: Black Hills Caverns, 2600 Cavern Road, Rapid City, SD 57702. Phone: (605) 343-0542. Web site: www.blackhillscaverns.com.

50 Sitting Bull Crystal Caverns
Pennington County

TOURING THIS CAVE is like "walking through a giant geode," according to a Sitting Bull Crystal Caverns brochure. In this case, the brochure is close to the mark. Much of the cave is lined with dogtooth spar crystals, some as long as 16 inches. Diamond Lake, a one-foot-deep pond, reflects the crystals covering the ceiling, and the effect is memorable—a cross between a city skyline at night and the mouth of some toothy, subterranean creature. There are also cave popcorn, flowstone, and other formations, but the dogtooth spar crystals are definitely the main attraction.

Getting to Diamond Lake requires more strenuous walking than is needed on the other Black Hills cave tours, but the trek—including the 162 steps down at the entrance and a ladder traversing one section—is worth the effort. The tour wends almost 800 feet

At Sitting Bull Crystal Caverns, visitors can grab a flashlight and explore two small caves on their own.

beneath the surface, and the 40-degree temperature is several degrees cooler than that of many nearby caves. Sitting Bull Caverns is also wetter than most others, and you can still see where mud washed in during the flood of June 1972, when a dam broke and water rushed through the area.

On the road to Mount Rushmore, this cave opened to the public before the famous monument was completed in 1941. To promote the cave, owners Alex and Mamie Duhamel and their son Bud, sought assistance from their friends the Lakota, who staged a Native American pageant at the site, providing an introduction to their culture and customs. A trading post offered an opportunity for the Lakota to acquire the goods they needed and to sell their arts and crafts.

The Duhamel's Lakota friends included Black Elk and Iron Hawk, who had survived the Battle of Little Big Horn in 1876. In fact, Black Elk named the caverns and pageant after Sitting Bull, to honor the holy man who had once camped near the cave and had predicted Sioux success at Little Big Horn.

Adjacent to the caverns' entrance, another opening beckons. Kids and adventurous parents can explore this tunnel on their own. It runs roughly 50 feet into the hill, ending in a small room.

The approach to the caverns is as adventurous as the cave tour. The mile-long drive from the highway features many narrow hairpin turns. Signs advise you to honk your horn as you approach each of the tight turns. Don't try this drive in a camper; and if you're pulling a trailer, you'll want to leave it at the parking lot near the highway.

Directions: From Rapid City, drive 9 miles south on U.S. Highway 16 to the cave road. Follow the incredibly twisting drive to the cave.

Season/hours: Open daily, 7 a.m. to 7 p.m., June through August. Open 8 a.m. to 5 p.m., May, September, and October.

Length: The tour takes about 45 minutes.

Precautions: To enter the cave you descend 162 steps, which you must climb on your return to the surface. The tour includes a ladder; be prepared to hold the railing with both hands to climb it safely.

Amenities: Gift shop includes Native American beadwork, jewelry, rugs, and other crafts.

Contact: Location: Sitting Bull Crystal Caverns, 13745 South U.S. 16, Rapid City, SD 57701. Mailing address: Sitting Bull Crystal Caverns, 6201 Magic Canyon Road, Rapid City, SD 57702.
Phone: (605) 342-2777.

Web site: www.sittingbullcrystalcave.com.
talcave.com.

51
Big Thunder Gold Mine
Pennington County

THE BIG THUNDER Gold Mine consists of a straight tunnel dug into the mountainside on the edge of the town of Keystone. Two German immigrants, W. B. Krupp and J. A. Engle, started this project in 1882, hoping to intersect a 10-foot-wide gold-bearing vein discovered by a neighboring mine.

A tour group enters Big Thunder Gold Mine in Keystone, South Dakota.

Initially the two miners dug into a layer of easily broken granite schist, so the first section of the tunnel is reinforced with timbers. In this timbered section, look for the powder room, where they stored their blasting powder. Krupp and Engle stored the powder near the entrance, so that in the event of an accidental explosion, area residents would see and hear the blast—and come to their rescue.

Beyond the schist, the miners encountered solid granite, which did not require timbering. In the granite section of the tunnel you walk by a sheet of iron about four feet square. As the miners worked, this iron sheet caught the rock that dropped from the ceiling and walls. The loud reverberations of the rock falling onto the iron sheet gave the name to the mine. (Hold your ears when the guide demonstrates by shaking and dropping the iron sheet.)

Eventually the miners, working mostly weekends, reached the gold-bearing vein of iron pyrite and graphite; however, at this point, the vein had narrowed to only one foot. When they followed it, it soon petered out altogether.

Not men to give up, the two miners kept digging, now aiming to intercept another strong vein recently discovered nearby. An investor offered financial help in exchange for 50 percent of any profits, and Krupp and Engle worked until they intercepted the new vein.

Once more, the effort turned out not to be worth the work. At the end of the tunnel the miners dug a 30-foot-high stope, taking out tons of ore. But the amount of gold was small: every seven tons of ore yielded only one ounce of gold. Following 32 years of mining at Big Thunder, Krupp and Engle split $100 in profits, after giving $100 to the investor who funded the extension of the tunnel.

Black Hills Mining Museum

For a more in-depth look at gold mining in the region, the Black Hills Mining Museum offers a realistic, hour-long tour of a simulated underground level of the Homestake Gold Mine, which ceased operations in 2001. Guides describe nineteenth-century and modern underground mining processes. More than 140 miners and former mine employees helped create this simulation.

Open year-round, the museum also features working models of mining equipment, historic photographs, a video theater, and gold panning. Check the museum's web site for information on special events like mining artifact auctions and mining contests—featuring hand mucking, drilling, and a track-tie toss.

Contact: P.O. Box 694, 323 W. Main Street, Lead, SD 57754. Phone: (605) 584-1605. Web site: www.mining-museum.blackhills.com.

Still, their effort today has paid off in the form of an attraction that offers an intriguing story, historic information, and a geology lesson. The tour ends with an explanation of old mining equipment and production processes.

Directions: From Rapid City, take U.S. Highway 16 for 18 miles to U.S. 16A. Drive 2 miles to State Highway 40 in Keystone. The mine is just east of town.

Season/hours: Open daily, 8 a.m. to 8 p.m., mid-June through August; open 9 a.m. to 5:30 p.m., May through mid-June and September through mid-October.

Length: The tunnel is 680 feet long. Allow an hour or two to tour the tunnel, explore old mining artifacts, and pan for gold.

Precautions: This is an easy tour for most individuals; it is entirely handicapped accessible. Some sections are no more than five-and-a-half-feet high.

Amenities: Old mining exhibits, gold panning lessons.

Contact: Big Thunder Gold Mine, P.O. Box 459, Keystone, SD 57751. Phone: (605) 666-4847, (800) 314-3917.

52

Thunderhead Underground Falls
Pennington County

A GLIMPSE INTO the unusual history of one mining operation awaits visitors to Thunderhead Underground Falls. In the 1870s, the mine, operated by the Fort Meade Hydraulic Gold Mining Company, grew into a tunnel dug into the side of a mountain. Using hand drills and black powder, miners slowly extended the tunnel, extracting ore as they went until ceasing all operations around 1900.

Exactly what happened then is undocumented. Local residents speculate that the miners blasted their way up and into the underground bed of Rapid Creek, which disappears underground into a sinkhole about a mile and a half away. The resulting waterfall ended the mining operation. Essentially, the company mined for gold and struck water.

Almost half a century later, two curious tourists, Albert and Vera

Eklund, discovered the tunnel with the waterfall at its end and bought the property from company heirs. They opened the attraction to the public in 1950.

The property has since changed hands again, but the current owners have continued to preserve the natural state of the canyon. There is no gift shop, picnic area, or gem mining attraction, just a small ticket office.

Your tour begins with a short audiotape presentation on the history of the tunnel. There are no guides—you couldn't hear them over the rush of water, and they are unnecessary anyway, because the passage is a straight trail into the mountain. You walk on a gravel path, with a wooden rail separating the path from the ditch carrying the rushing runoff from the falls. Signs along the way point out hand-drill marks, gold-bearing quartz veins, and small stalactites that have begun to grow from the ceiling.

The roar of the 30-foot falls increases near the end of the tunnel. A short roof near the falls provides some protection from the spray, although a mist permeates much of the tunnel.

While this tour is a bit primitive, lacking guides and cement walkways, it is also easier than most of the area's other underground tours. Most children and adults have no problem walking the 600 feet along the flat path. And the oddity at the tunnel's end provides a rewarding spectacle and an intriguing glimpse into the challenges of early mining.

Directions: From Rapid City, drive 10 miles west on State Highway 44 to the drive leading to the falls.

Season/hours: Open daily, 8 a.m. to dusk, May through October.

Length: The adit runs 600 feet into the mountain.

Precautions: The tunnel is usually filled with a heavy mist from the falls.

Amenities: An old ore train is on display at the site.

Contact: Phone: (605) 343-0081.

53

Beautiful Rushmore Cave

Pennington County

ONE OF THE most visited of the Black Hills caves, Beautiful Rushmore Cave, named for its proximity to Mount Rushmore, is the most accessible of the region's commercial caves. A tour still includes many stairs, one ladder, and a few narrow passages where adults must stoop; but there are also easy walking passages and the man-made exit tunnel is a gentle ramp to the gift shop. Because many people visit and warm this cave, its temperature is relatively high—up to 58 degrees on a busy summer day.

This entrance building/gift shop perches part way up a bluff near the cave opening. Leave your car in the parking lot and climb broad wood steps to the building, where you can catch your breath, browse, and buy snacks and tour tickets.

Gold miners discovered the entrance to this cave when their flume broke and water disappeared into the ground. Even 100 years ago, visitors explored the cave, and some carved their names in one section, since dubbed the Post Office. Today's tour passes this early graffiti.

Beyond the Post Office and the Image Room, the Big Room offers the most diverse collection of speleothems and is the turning point of your tour. Stalactites, stalagmites, flowstone, ribbon, draperies, columns, and helictites decorate the walls and ceiling.

Visitors often photograph the formations in this room, and carrying a bright flashlight can sometimes help you illuminate the speleothems you're trying to capture on film. Although the commercial caves are all wired and lighted, there are many obscure corners and formations.

After tracing your steps to the Image Room, you follow a different passage on your return trip. This route leads to the picturesque Floral Room and eventually to the exit tunnel. Your guide may point out the side passage that leads to Mystic Cave, one of many wild caves in the region.

Directions: **From Rapid City, take U.S. Highway 16 for 18 miles to U.S. 16A. Drive 2 miles to State Highway 40 in Keystone. From Keystone, drive 6 miles east on 40 to the cave.**

Season/hours: **Open daily, 8 a.m. to 8 p.m., June through August. Open 9 a.m. to 5 p.m., May, September, and**

October.

Length: The .5 mile tour takes about one hour.

Precautions: This cave is partially wheelchair accessible, but the regular tour contains several passages too narrow for wheelchair access.

Amenities: A written version of the tour is available for people with hearing disabilities.

Contact: Beautiful Rushmore Cave, 13622 Highway 40, Keystone, SD 57751. Phone: (605) 255-4384 or (605) 255-4467. Web site: www.beautifulrushmorecave.com.

54

Custer State Park

Custer County

Custer State Park's Hole-in-the-Wall Picnic Area, on the route to the spiky spires of the Needles Scenic Highway, is the only easily accessible cave in the park. From the parking area, walk across a short bridge to reach this small cave at the base of a bluff.

Rock climbers can find small cavities in a few cliffs, but the numerous attractions of Custer State Park—principally, buffalo, elk, pronghorn, mountain goats, bighorn sheep, coyotes, prairie dogs, whitetail and mule deer, as well as the tunnels—do not include caves.

Mining, however, played a major role in the development of the park. An expedition led by Lieutenant Colonel George Armstrong Custer discovered gold along the banks of French Creek in the summer of 1874. The following winter, a group of gold seekers, the Gordon Party, came to the region from Iowa, building a stockade as a protection against attack from the Lakota, who owned the land under the 1868 Treaty of Fort Laramie. Gordon Party members realized little success from their gold panning efforts and within five months of their arrival at French Creek, the U.S. Cavalry moved them from the Hills. But word of the find had gotten out and within a year almost 20,000 gold seekers—more than the army could remove—were on the scene. Still, no one ever much profited from the gold in French Creek. Today, a replica of the Gordon Stockade marks the original site.

The park does, of course, contain other underground attractions: two of the five tunnels along the Peter Norbeck Scenic Byway are within the Park. (See page 154.)

The highlight of Custer State Park's Hole-in-the Wall Picnic Area is this small cave.

Directions: From the Custer State Park Office, follow U.S. Highway 16A west to State Highway 87. Turn right and follow 87 to County Highway 753 (road to Black Hills Playhouse). Follow 87 as it turns left. The Hole-in-the-Wall Picnic Area is roughly 4 miles ahead on the left. (This section of 87 is known as the Needles Highway Scenic Drive.)

Season/hours: Open year-round, weather permitting.

Precautions: Parents will want to keep an eye on small children on the bridge and stream banks.

Amenities: Custer State Park offers activities for every visitor—camping, hiking, panning for gold, picnicking, and rock climbing.

Contact: Custer State Park, HC 83, Box 70, Custer, SD 57730. Phone: (605) 255-4515. Web site: www.custerstatepark.info/index.htm.

55

Peter Norbeck Scenic Byway

Custer County

THE PETER NORBECK Scenic Byway encompasses the Iron Mountain Road and the Needles Highway, providing the opportunity to drive through five outstanding tunnels. On a clear day, three offer a view of Mount Rushmore, and two traverse the Needles of the Black Hills. All are on paved roads. They vary in width from less than 9 feet to 14 feet. None is more than a few car lengths deep.

Peter Norbeck, a South Dakota governor and United States senator, laid out these roads on foot and horseback, to best preserve the natural features of the Harney Range. At its highest elevations, the Range exposes granite spires, columns, and outcrops. At lower elevations, streams flow through meadows and ponderosa pines, aspen, and oak groves.

Directions: To the Iron Mountain Road tunnels: From the Keystone, follow U.S. 16A (Iron Mountain Road) south approximately 5 miles to the three tunnels. If you drive through them, turn around, and then approach the tunnels from the south, each frames Mount Rushmore in the distance. To the Needles Highway tunnels: From Keystone, follow U.S. 16A (Iron Mountain Road) south through the three tunnels. Continue on 16A as it enters Custer State Park and turns west. Turn right on State Highway 87 (Needles Highway) and follow it about 14 miles to the Needles and the two tunnels through these granite columns.

Season/hours: Due to the sharp curves and steep grades, sections of the byway are closed during the winter.

Length: None of the tunnels are more than a few car lengths deep.

Precautions: Watch your speed. Oncoming cars, bicyclists, mountain goats, and other animals may be just around the next twist in the road. The lowest tunnel measures only 10 feet, 8 inches high.

Amenities: Several pull-offs offer scenic views, and a few picnic areas provide space for a longer break.

Contact: Peter Norbeck Visitor Center, Custer State Park, HC 83, Box 70, Custer, SD 57730. Phone: (605) 255-4515. Web site: www.byways.org.

This is one of two tunnels through the Needles in Custer State Park.

56
Jewel Cave National Monument
Custer County

CONTAINING CLOSE TO 130 miles of mapped passages, Jewel Cave is the third largest in the world and the second largest in the United States. (At more than 350 miles, Kentucky's Mammoth Cave is currently the longest.) Yet many scientists believe that hundreds of miles of undiscovered passages lie hidden in Jewel Cave.

Discovered about 1900, Jewel Cave was believed to be less than a half mile long for almost 60 years. Two brothers, Frank and Albert Michaud, and their friend Charles Bush found the cave while riding through Hell Canyon, where their horses were spooked by a whistling noise. Investigating, they found a breeze issuing from a 14-inch hole in the earth. On that day the cave was exhaling, since the air pressure inside was lower than the pressure outside.

The brothers and their friend enlarged the hole and entered the cave. They found many calcite crystals, which sparkled like jewels in their lantern light. The Michaud brothers, who were prospectors, at first believed they could mine the crystals and actually filed a claim on the Jewel Tunnel Lode in Custer. The crystals, however, turned out to be worthless outside of the highly decorated cave environment.

Plan B for the Michauds and Bush was to develop the cave as a tourist attraction. They developed a trail inside the cave and built a lodge on the canyon rim. But travel in the region was difficult in the early twentieth century, and their endeavor proved unprofitable. Still, the Michauds labored on, buying out Bush and maintaining the cave.

Slowly word spread of the cave's multitude of crystals and formations, and in 1908 President Theodore Roosevelt declared the cave a national monument. With assistant from area businessmen, the Civilian Conservation Corps, and the National Park Service, Jewel Cave was developed in subsequent decades, with the addition of several access roads, an office building, a stone stairway, and other improvements.

The biggest assistance to the cave's development came in 1959 in the form of two rock climbers recruited to help explore new underground passages. Geologist Dwight Deal invited Herb and Jan Conn to investigate and map the cave. Initially less than enthusiastic about climbing into dark, dirty holes, the Conns quickly became enthralled

Crystals coat many of the walls in the passages and chambers of Jewel Cave.

by the cave's mysteries and committed to finding new passages. They went on to lead more than 700 trips into Jewel Cave, spending some 6,000 hours underground. By the time they retired in 1980, the Conns had added more than 60 miles to Jewel's mapped passageways. Every year since, new generations of cave explorers have discovered more of the extensive cave system.

You now can see the sparkling speleothems, passages coated with crystals like wallpaper, and a fraction of this mazelike cave system on three distinct tours offered by National Park Service rangers.

- The Scenic Tour includes the Formation Room and other chambers highly decorated with nail-head spar crystals, stalactites, stalagmites, flowstone, draperies, an impressive 20-foot cave bacon formation, and even a soda straw column. This tour takes about 75 minutes and covers a half-mile route. An elevator ride in and out of the cave eases the challenge, but this tour still includes 723 stairs.

- The Candlelight Tour begins at the cave's historic entrance and includes steep stairs and more stooping and bending than does the Scenic Tour. The Candlelight Tour takes about 105 minutes. You carry a candle-lantern on this trip, which is open to children ages six and older.
- The Spelunking Tour is only for adventurous adults and older teens (16- and 17-year-olds need parental permission), who must first qualify by squirming through a tunnel measuring 8.5 by 24 inches. This .5-mile tour takes four hours.

Jewel Cave welcomes thousands of visitors each year, and the National Park Service has taken many precautions to preserve this national treasure. Lint tarps under the metal mesh walkways on the Scenic Tour collect more than 100 pounds of lint annually. Strategic lighting helps prevent the growth of moss, and tour group size is limited. During the busy summer season, call ahead for reservations or reserve your place online.

Directions: From Custer, follow U.S. Highway 16 west for 13 miles to the cave.

Season/hours: Open daily, 8 a.m. to 7:30 p.m., during the summer. Visitor Center is open year-round, generally from 8 a.m. to 5 p.m., except on Thanksgiving Day, Christmas Day, and New Year's Day.

Length: The cave is more than 127 miles long. Guided tours range from about .5 to .7 miles.

Precautions: Reservations required for some of the cave tours. Cave temperature averages 49 degrees. The visitor center and Target Room in the cave are wheelchair accessible.

Amenities: Educational exhibits, hiking trails, summer interpretative programs.

Contact: Jewel Cave National Monument, RR 1, Box 60 AA, Custer, SD 57730. Phone: (605) 673-2288. Web site: www.nps.gov/jeca.

57
Wind Cave National Park
Custer County

THE LENGTH OF Wind Cave's mapped passages grows annually, as cave explorers continue to check new leads and follow air currents into unknown areas. With 103 miles of surveyed passages, Wind Cave is the sixth longest cave in the world and contains 95 percent of the world's known boxwork. Flowstone, stalactites, stalagmites, popcorn, frostwork, and helictite bushes are also on display throughout Wind Cave; solution pockets are also visible on the ceiling, where water swirled during the cave's development.

Wind Cave was discovered by Jesse and Tom Bingham in 1881, although Native Americans may have known of its existence earlier. While deer hunting, the brothers heard a whistling noise, investigated, and came across an 8-inch hole in the ground. Story has it that the wind was blowing so hard, Jesse's hat blew off. Even today, the wind whistles in and out of the small opening at speeds up to 80 miles per

A group explores Wind Cave via the Natural Entrance Tour.

hour. The wind rushes in or out to equalize the air pressure between the cave and the above-ground environment.

Credit for discovering the cave's fascinating formations and many of its mazelike passages goes to Alvin McDonald, who was just 17 when he first squeezed through the narrow opening and began systemically recording his findings. Soon, the McDonald family and other local entrepreneurs blasted open a larger entrance, widened the smaller passages, and began offering guided tours into the cave. In 1903, Wind Cave was named a national park.

Today, you can still see the tiny opening that McDonald pushed through on the Natural Entrance Tour, one of five tours guided by National Park Service staff:

1. Garden of Eden Tour—the easiest tour available visits the Garden of Eden Room and showcases many different formations. You enter and leave the cave by elevator and face only 150 stairs on the one-hour tour.
2. Natural Entrance Tour—this trip to the middle level of the cave, about 220 feet beneath the surface, features abundant boxwork. You walk into the cave near the natural opening and leave by elevator. This tour lasts 75 minutes and includes 300 stairs, mostly descending. The last stop in the cave is the Assembly Room, with more than half a dozen passages radiating from it, like the spokes in a wheel. These openings offer good frames for photos.
3. Fairgrounds Tour—this moderately challenging tour has 450 stairs and leads to some of the larger rooms in the cave. It showcases boxwork and other speleothems and takes 90 minutes.
4. Candlelight Tour—this trip mimics the cave tours conducted a century ago, when visitors carried candles to light their way. (You carry a candle-bucket.) This strenuous tour covers one mile of rough trail and takes two hours. Required: shoes with nonslip soles, no sandals. Minimum age is eight.
5. Wild Cave Tour—this introduction to safe caving is a four-hour tour, with much crawling. Hard hats, lights, and kneepads are supplied, but you need to wear old clothes (long pants and long sleeves), boots or nonslip shoes, and gloves. Minimum age is 16; 16- and 17-year-olds must have a signed consent form.

The number of visitors on each tour is limited, from 10 on the Candlelight and Wild Cave Tours to 40 on the easier tours. Reservations are required for the Wild Cave Tours and recommended for the other tours, particularly during the summer. More than 150,000 visitors tour the cave each year.

In the face of this popularity, the National Park Service is taking precautions to preserve Wind Cave. One concern is that its lakes, 400 feet below the surface, are part of the region's aquifer. A current experiment involves collecting water dripping into the cave and conducting water quality tests. Part of the cave system lies beneath the visitor parking lot, so dirt, oil, antifreeze, and other elements will eventually leach into the cave. The Park Service is considering reconstructing the lot and installing a filtering system to preserve water quality in the cave—and in the local aquifer.

Directions: From Rapid City, follow State Highway 79 south for 50 miles to U.S. Highway 385. Turn west on 385 and drive for 5 miles to Hot Springs. From Hot Springs, continue north on 385 for 13 miles to the park. Watch for signs for the visitor center.

Season/hours: Open year-round; visitor center open daily except Thanksgiving and Christmas. From early October to early April, open 8 a.m. to 4:30 p.m.; from early April to early May, 8 a.m. to 5 p.m.; from early May to mid-June, 8 a.m. to 6 p.m.; from mid-June to mid-August, 8 a.m. to 7 p.m.; from mid-August to mid-September, 8 a.m. to 6 p.m.; from mid-September to early October, 8 a.m. to 5 p.m.

Length: More than 103 miles of passages have been mapped so far.

Precautions: Some sections of the tour paths may be wet and slippery. Temperature is 53 degrees, so a light jacket or sweatshirt is recommended.

Amenities: Above ground, the park's 28,295 acres of prairie and ponderosa pine forest provide habitat for buffalo, elk, pronghorn, mule deer, coyotes, and prairie dogs. Camping, picnicking, hiking, and summer interpretative programs are offered.

Contact: Wind Cave National Park, RR1 Box 190, Hot Springs, SD 57747. Phone: (605) 745-4600. Web site: www.nps.gov/wica.

Appendix A
Sites by Type

Natural Caves, Arches, and Fissures

1. Niagara Cave
2. Mystery Cave I
3. Mystery Cave II
4. Forestville State Park
5. Cave Farm Karst Preserve
6. Masonic Park
7. Bear Cave Park
9. Whitewater State Park
10. Skunk Hollow Cave
11. Winona Bluffs
12. Garvin Heights
13. La Moille Cave
14. Pipestone National Monument
15. Blue Mounds State Park
16. Wyalusing State Park
19. Wabasha Street Caves
21. Joseph Wolf Caves
22. Interstate State Park
23. Crystal Cave
25. Small cave on approach to Robinson's Ice Cave
38. Gooseberry Falls State Park
39. North Shore Sea Caves
41. Isle Royale National Park
45. Bethlehem Cave
46. Beautiful Wonderland Cave
47. Stagebarn Crystal Cave
48. Crystal Cave Park
49. Black Hills Caverns
50. Sitting Bull Crystal Caverns
53. Beautiful Rushmore Cave
54. Custer State Park
56. Jewel Cave National Monument
57. Wind Cave National Park

Manmade Caves and Tunnels

8. Quarry Hill Park
17. Carlson Lime Kiln
18. Red Wing Memorial Park
36. Ely's Peak Tunnel
37. Highway 61 Tunnels
43. George S. Mickelson Trail
55. Peter Norbeck Scenic Byway

Pit Mines

27. Hill Annex Mine
28. Hull Rust Mahoning Mine Overlook
29. Ironworld Discovery Center
30. Mountain Iron Historic Overlook
31. Leonidas Overlook
32. Mineview in the Sky
34. Wacootah and Minntac Overlooks
40. Voyageurs National Park

Underground Mines

35. Soudan Underground Mine State Park
44. Broken Boot Gold Mine
51. Big Thunder Gold Mine
52. Thunderhead Underground Falls

Miscellaneous Sites

20. Fort Snelling State Park
24. Banning State Park
26. Croft Mine Historical Park
33. Minntac Taconite Plant
42. Amethyst Mine Panorama

Appendix B
Resources

Books

Alexander, E. Calvin, ed. *An Introduction to Caves of Minnesota, Iowa, and Wisconsin.* White Bear Lake, Minn.: Lakewood Community College, 1980. Published for the 1980 National Speleological Society Convention, this volume discusses caves both accessible and inaccessible to the public. It can be ordered from the Minnesota Geological Society.

Hogberg, R. K., and T. N. Bayer. *Guide to the Caves of Minnesota.* Minneapolis: University of Minnesota, 1967. This volume is number four of an educational series produced by the Minnesota Geological Survey. While this book contains much valuable background, it also contains several errors—for example, stating that La Moille Cave was destroyed by highway construction.

Park, John R. *Guidebook to Mining in America.* South Miami, Fla.: Stonerose Publishing Company, 2000.

Perry, David E. *Gold Town to Ghost Town: Boom and Bust on Rainy Lake.* International Falls, Minn.: Lake States Interpretative Association, 1993.

Runkel, Anthony C. *The Geology of Whitewater State Park.* St. Paul: University of Minnesota, 1996. This volume is number nine of an educational series produced by the Minnesota Geological Survey.

Sansome, Constance J. *Minnesota Underfoot: A Field Guide to the State's Outstanding Geologic Features.* Bloomington, Minn.: Voyageur Press, 1983.

Sims, P. K., and G. B. Morey. *Geologic Sketch of the Tower-Soudan State Park.* Minneapolis: University of Minnesota, 1966. This volume is number three of an educational series produced by the Minnesota Geological Survey.

Tomovick, Barbara, and Kimberly Metz. *The Insiders' Guide to South Dakota's Black Hills and Badlands.* Guilford, Conn.: Insiders' Publishing, Inc., 2000.

Organizations

Minnesota Geological Survey, 2642 University Avenue W., St. Paul, MN 55114-1057. Phone: 612-627-4780. Web site: www.geo.umn.edu/mgs/.

Minnesota Historical Society, 345 W. Kellogg Boulevard, St. Paul, MN 55102-1906. Phone: (651) 296-6126. Web site: www.mnhs.org. Minnesota Speleological Survey (MSS), e-mail: mss@intothevoid.com. Web site: www.intothevoid.com/mss/. You may also learn about the MSS by attending a monthly meeting. Newcomers are welcome.

Web Sites

http://cavern.com. The National Caves Association is a nonprofit organization of member commercial caves, and its site provides a list of commercial caves across the United States.

www.caves.org. The National Speleological Society offers memberships, publications, videos, events, and links to cave-related sites.

www.dnr.state.mn.us/parks_and_recreation/state_parks/. The Minnesota Department of Natural Resources provides information on state parks, trails, campgrounds, and other outdoor attractions.

www.dnr.state.mn.us/minerals/digging_minerals/history.html. This site offers background on minerals, mining, fossils, and other geology topics.

www.geo.umn.edu/mgs/. The Minnesota Geological Survey develops maps and educational materials. Its site also offers The Virtual Egg Carton, a description of common state rocks.

www.geocities.com/urbanspeleology. Geologist Greg Brick introduces visitors to underground sites in the Twin Cities and provides information and links to selected urban sites around the nation.

http://home.att.net/~ackerman.john/. John Ackerman's site describes the former commercial cave, Spring Valley Caverns, and the development of Cave Farm Karst Preserve. There's also information on upcoming events at the preserve.

www.intothevoid.com/mss/. The Minnesota Speleological Survey welcomes new members, and its site offers information on meetings, membership, and projects, as well as a list of essential caving supplies.

http://irontrail.org. The Iron Trail Convention and Visitors Bureau provides information on activities, lodging, and maps for the cities of the Mesabi Iron Range.

www.iteccomputers.com/iron_range_links.html. Search this site for a variety of Iron Range links.

http://karst.wku.edu/. Western Kentucky University's Karst Studies Group explores a variety of research projects and lists upcoming conferences on karst scholarship. The University is located in Bowling Green, in the south-central Kentucky karst region.

www.mnhs.org. The Minnesota Historical Society provides a wealth of historic information on mines, quarries, and other sites.

www.nps.gov/. The National Park Service provides links to Wind Cave National Park and other properties featuring caves, as well as information on their geology, history, and development.

www.preservecampcoldwater.org. The Preserve Camp Coldwater Coalition seeks to protect and restore historic Coldwater Spring.

www.rangecities.com/. This site offers information on the history and current attractions of more than two dozen communities in the state's three Iron Ranges.

www.state.sd.us/gfp/sdparks/listing.htm. The South Dakota State Park and Recreation Office links you to Custer State Park, the George S. Mickelson Trail, and many other sites of interest.

www.tc.umn.edu/~smith213/index.htm. The Geological Society of Minnesota is a nonprofit membership organization offering lectures, field trips, and publications. Members can purchase for the cost of postage a St. Cloud granite and mineral collection.

www.wildnorth.org. This site provides visitor information for the communities of the Cuyuna Iron Range.

Glossary
of Caving and Mining Terms

adit—a horizontal passage into a mine.
agglomerator—machine in which magnetite concentrate is mixed with a binding agent, formed into marble-sized taconite pellets, and baked.

bacon onyx—a formation of onyx in a solution cave that resembles strips of bacon hanging sideways from the ceiling or walls.
big brown bat—Minnesota's largest species of cave bat.
blind valley—dead-end ravine; box canyon.
boxwork—a calcite formation resembling a honeycomb grown wild, with many compartments of angular, irregular shapes.

calcite—calcium carbonate, a mineral found in the form of limestone, with hexagonal crystals.
Cenozoic era—the current time of the Phanerozoic eon in which mammals have dominated the earth.
cephalopod—a marine animal belonging to the same family as octopi and squid.
column—in a cave, a formation created when a stalactite and stalagmite meet and grow together.

dolostone—a rock composed of dolomite, a mineral containing calcium and magnesium carbonate.
drift—(1) rock and other material transported and left behind by glaciers; (2) horizontal passage in a mine, along a mineral vein.
driftless area—a region not covered by glaciers and free of glacier drift, such as southeastern Minnesota.

eastern pipistrelle—one of four types of cave bats found in Minnesota.
escarpment (or scarp)—a cliff or steep drop-off, for example, at the edge of a plateau.

flowstone—smooth-surfaced mineral buildup in a solution cave that resembles an ice layer.
fossils—remains of once-living organisms or their evidence, for example, footprints or casts, preserved in rock.

goethite—an iron-bearing major ore mineral named after Johann Wolfgang von Goethe, author of Faust.
Granby car—a side-dumping mine car that miners used to transport ore in a mine.

headframe—a structure above a mine shaft that supports the hoisting gear. See "hoist house."
helictite—a hook-, twig- or bushlike formation created when water seeps slowly out of the side of a stalactite instead of dripping straight off the end.
hematite—iron oxide, a major ore mineral.
Hoist—an engine, along with cables and related apparatus, that lifts people and materials in and out of a mineshaft.
hoist house—a shelter for the winch and engine used to lift people and materials from a mine shaft or lower them into it.

illite–a general term for micalike clay minerals often found with shales. Clays containing illite are used extensively for makeup, since they absorb facial oils.

karst—landscape featuring sinkholes, springs, and disappearing streams above ground and cave systems beneath the surface.

lime kiln—an oven in which limestone is heated to produce lime.
limestone—rock comprised mostly of calcite (calcium carbonate).
little brown myotis—a common cave bat in Minnesota.
loaf—large block of granite, separated from the rock layer in a quarry.
location—a mining camp or town, often constructed for the workers by the company.
lumper—unskilled quarry worker.

manrise—steeply inclined or vertical passages connecting drifts and subdrifts in a mine.
manway—a passage enabling miners to move from place to place in a mine, for example, in a manrise.
Mesozoic era—the period in the Phanerozoic eon from about 230 million years to about 65 million years ago, including the time when dinosaurs dominated the planet. In Minnesota, lands uplifted, and wind and water eroded the exposed rock.
moraine—a mound, hill, or other mass of glacial drift, containing boulders, gravel, clay, and sand.

northern myotis—one of four species of Minnesota cave bats.

onyx—a variety of chalcedony having an almost waxy appearance and bands of different colors.
Ordovician period—an early time frame of the Paleozoic era. Many Minnesota caves are formed in rock layers of this period.
ore—rocks composed of ore minerals from which useful materials can be extracted.

Paleozoic era—the period ending about 230 million years ago. Shallow seas covered much of the state during this era, which belongs to the Phanerozoic eon.
petroglyph—early rock-art carving, engraving.
Phanerozoic eon—from 570 million years ago to today; the time when sea life became abundant, the first vertebrates appeared, and, finally, life became widespread.
Pleistocene epoch—age of the glaciers, 10,000 years ago.
Precambrian eon—more than 570 million years ago when primitive one-celled and multi-celled organisms first appeared.

quartz—hard mineral composed of silica.
Quaternary period—the most recent period of the current Cenozoic era.

sandstone—a sedimentary rock composed of sand grains cemented by silica or other materials.
schist—a metamorphic rock (formed through many changes in temperature, pressure, and environment) composed of layers that sometimes split into platelike fragments.
Scraper bucket—an implement that was dragged across the floor of a mine to rake the ore toward a cart for transport out of the mine.
sea cave—cavity in a shoreline bluff hollowed out by wave erosion.
shaft—a vertical or steeply inclined passageway into the earth.
shaft house—a shelter constructed above a mineshaft.
shale—a fine-grained rock formed from clay.
sinkhole (or sink)—a depression or hole in the earth's surface, sometimes leading to a cave.
skip—a mine car that rolled on a track and was used to carry ore from a mine to the surface.
slab—to cut a slice from a block of granite.
soda straws—incipient stalactites that form as hollow, circular mineral deposits growing from the ceiling of solution caves.
solution cave—an underground cavity formed primarily by the dissolution of a soluble rock.

speleothem—cave formation such as a stalactite or stalagmite.
stalactite—a cave formation that grows downward from the ceiling like a lengthening and thickening icicle.
stalagmite—a cave formation that grows upward from the floor like a lengthening and thickening cone.
stope—an underground chamber from which ore has been extracted. Can also refer to the rock face from which ore is excavated and be used as a verb meaning to excavate upward.
supermodule—a machine containing many wide steel plates, used in an experiment to detect whether neutrinos have changed type in the Main Injector Neutrino Oscillation Search (MINOS) project located at Soudan Underground Mine.

taconite—iron-bearing major ore mineral containing magnetite, as well as hematite.
tailings—fine-grained waste from ore mills.
talus—pile of rocks at the base of a slope or bluff.
till—glacial debris including soil and rocks.
tugger—a winch powered by an integral engine and used to move the scraper bucket in a mine.

vug—a small cavity in a rock—for example, in the wall of a cave passage—that is lined with crystals, similar in appearance to a geode.

Index

MORE GREAT TITLES FROM TRAILS BOOKS & PRAIRIE OAK PRESS

ACTIVITY GUIDES

Great Cross-country Ski Trails: Wisconsin, Minnesota, Michigan & Ontario, *Wm. Chad McGrath*

Great Minnesota Walks: 49 Strolls, Rambles, Hikes, and Treks *Wm. Chad McGrath*

Great Wisconsin Walks: 45 Strolls, Rambles, Hikes, and Treks *Wm. Chad McGrath*

Acorn Guide to Northwest Wisconsin, *Tim Bewer*

Paddling Illinois: 64 Great Trips by Canoe and Kayak, *Mike Svob*

Paddling Southern Wisconsin: 82 Great Trips by Canoe and Kayak *Mike Svob*

Paddling Northern Wisconsin: 82 Great Trips by Canoe and Kayak *Mike Svob*

Wisconsin Underground: A Guide to Caves, Mines, and Tunnels in and around the Badger State, *Doris Green*

TRAVEL GUIDES

Great Little Museums of the Midwest, *Christine des Garennes*

Great Minnesota Weekend Adventures, *Beth Gauper*

The Great Wisconsin Touring Book: 30 Spectacular Auto Tours *Gary Knowles*

Tastes of Minnesota: A Food Lover's Tour, *Donna Tabbert Long*

Wisconsin Lighthouses: A Photographic and Historical Guide *Ken and Barb Wardius*

Wisconsin Waterfalls, *Patrick Lisi*

Wisconsin Family Weekends: 20 Fun Trips for You and the Kids *Susan Lampert Smith*

County Parks of Wisconsin, ***Revised Edition,*** *Jeannette and Chet Bell*

Up North Wisconsin: A Region for All Seasons, *Sharyn Alden*

Great Wisconsin Taverns: 101 Distinctive Badger Bars, *Dennis Boyer*

Great Weekend Adventures, *the Editors of Wisconsin Trails*

The Wisconsin Traveler's Companion: A Guide to Country Sights, *Jerry Apps and Julie Sutter-Blair*

NATURE ESSAYS

Wild Wisconsin Notebook, *James Buchholz*

Trout Friends, *Bill Stokes*

Northern Passages: Reflections from Lake Superior Country
Michael Van Stappen

River Stories: Growing up on the Wisconsin, *Delores Chamberlain*

HOME & GARDEN

Creating a Perennial Garden in the Midwest, *Joan Severa*

Wisconsin Garden Guide, *Jerry Minnich*

Wisconsin Herb Cookbook
Suzanne Breckenridge & Marjorie Snyder

Bountiful Wisconsin: 110 Favorite Recipes, *Terese Allen*

Wisconsin's Hometown Flavors: A Cook's Tour, *Terese Allen*

HISTORICAL BOOKS

Prairie Whistles: Tales of Midwest Railroading, *Dennis Boyer*

Barns of Wisconsin, *Jerry Apps*

Portrait of the Past: A Photographic Journey Through Wisconsin 1865-1920, *Howard Mead, Jill Dean, and Susan Smith*

Wisconsin: The Story of the Badger State, *Norman K. Risjord*

Wisconsin At War: 20th Century Conflicts Through the Eyes of Veterans, *James F. McIntosh, M.D.*

GIFT BOOKS

The Spirit of Door County: A Photographic Essay, *Darryl R. Beers*

Milwaukee, *Photography by Todd Dacquisto*

Duck Hunting on the Fox: Hunting and Decoy-Carving Traditions
Stephen M. Miller

Spirit of the North: A Photographic Journey Through Northern Wisconsin, *Richard Hamilton Smith*

GHOST STORIES

Haunted Wisconsin, *Michael Norman and Beth Scott*

W-Files: True Reports of Wisconsin's Unexplained Phenomena
Jay Rath

Northern Frights: A Supernatural Ecology of the Wisconsin Headwaters, *Dennis Boyer*

Giants in the Land: Folktales and Legends of Wisconsin,
Dennis Boyer

For a free catalog, phone, write, or e-mail us.

Trails Books
P.O. Box 317, Black Earth, WI 53515
(800) 236-8088 • e-mail: books@wistrails.com
www.trailsbooks.com